T0174705

"Western commentators tend to ignore the impact of global warming on 'developing' nations. Jahnnabi Das' study of the similarities and differences between Bangladeshi and Australian newspaper reporting of climate change might embarrass them. A pioneering and unexpected contribution to the global debate."

–Philip Bell, Emeritus Professor,
UNSW, Australia

"*Reporting Climate Change in the Global North and South* is a focused and informed examination of the similarities and differences in climate change communication between two vastly dissimilar countries. It is a theoretically informed analysis of how print media in Australia and Bangladesh deal with the vital question of climate change. This book intelligently traces elements of political cosmopolitanism emanating from two discursive positions – economic rationality and ecological vulnerability. This skillfully presented book is a valuable contribution to the comparative communication studies."

–Leslie Sklair, Emeritus Professor,
London School of Economics, UK

"*Reporting Climate Change in the Global North and South* breaks new ground in the analysis of climate communication. It shows how climate is a key issue for journalists in both Northern and Southern contexts. It finds similar dynamics in both Bangladesh and Australia, highlighting how climate reporting is fast developing as a nationally-embedded global discourse. The framing of climate news is a critical issue and this book makes a vital contribution to understanding it."

–James Goodman, Professor and Director,
Climate Justice Research Centre,
University of Technology Sydney, Australia

Reporting Climate Change in the Global North and South

This book reveals how journalists in the global North and global South mediate climate change by examining journalism and reporting in Australia and Bangladesh. This dual analysis presents a unique opportunity to examine the impacts of media and communication in two contrasting countries (in terms of economy, income and population size) which both face serious climate change challenges.

In reporting on these challenges, journalism as a political, institutional and cultural practice has a significant role to play. It is influential in building public knowledge and contributes to knowledge production and dialogue; however, the question of who gets to speak and who doesn't is a significant determinant of journalists' capacity to establish authority and assign cultural meaning to realities. By measuring the visibility from presences and absences, the book explores the extent to which the influences are similar or different in the two countries, contrasting how journalists' communication power conditions public thought on climate change. The investigation of climate communication across the North–South divide is especially urgent given the global commitment to reduce greenhouse gas emissions, and it is critical we gain a fuller understanding of the dynamics of climate communication in low-emitting, low-income countries as much as in the high-emitting, high-income countries. This book contributes to this understanding and highlights the value of a dual analysis in being able to draw out parallels, as well as divergences, which will directly assist in developing cross-national strategies to help address the mounting challenge of climate change.

This book will be of great interest to students and scholars of climate change and environmental journalism, as well as media and communication studies more broadly.

Jahnnabi Das is a Research Associate in the Climate Justice Research Centre at the University of Technology Sydney, Australia.

Routledge Studies in Environmental Communication and Media

For more information about this series, please visit: https://www.routledge.com/Routledge-Studies-in-Environmental-Communication-and-Media/book-series/RSECM

Reporting Climate Change in the Global North and South
Journalism in Australia and Bangladesh

Jahnnabi Das

LONDON AND NEW YORK

First published 2019
by Routledge
2 Park Square, Milton Park, Abingdon, Oxon OX14 4RN

and by Routledge
605 Third Avenue, New York, NY 10017

First issued in paperback 2021

Routledge is an imprint of the Taylor & Francis Group, an informa business

© 2019 Jahnnabi Das

The right of Jahnnabi Das to be identified as author of this work has been asserted by him in accordance with sections 77 and 78 of the Copyright, Designs and Patents Act 1988.

All rights reserved. No part of this book may be reprinted or reproduced or utilised in any form or by any electronic, mechanical, or other means, now known or hereafter invented, including photocopying and recording, or in any information storage or retrieval system, without permission in writing from the publishers.

Trademark notice: Product or corporate names may be trademarks or registered trademarks, and are used only for identification and explanation without intent to infringe.

British Library Cataloguing-in-Publication Data
A catalogue record for this book is available from the British Library

Library of Congress Cataloging-in-Publication Data
Names: Das, Jahnnabi, author.
Title: Reporting climate change in the global North and South : journalism in Australia and Bangladesh / Jahnnabi Das.
Description: London ; New York, NY : Routledge, 2019. |
Series: Routledge studies in environmental communication and media |
Includes bibliographical references and index. | Description based on print version record and CIP data provided by publisher; resource not viewed.
Identifiers: LCCN 2019009955 (print) | LCCN 2019014431 (ebook) |
ISBN 9780429402210 (ebook) | ISBN 9781138392403 (hardback)
Subjects: LCSH: Climatic changes--Press coverage. | Climatic changes--Press coverage--Australia. | Climatic changes--Press coverage--Bangladesh.
Classification: LCC PN4784.C624 (ebook) | LCC PN4784.C624 D37 2019 (print) | DDC 070.4/4955163--dc23
LC record available at https://lccn.loc.gov/2019009955

ISBN 13: 978-0-367-77768-5 (pbk)
ISBN 13: 978-1-138-39240-3 (hbk)

Typeset in Times New Roman
by Taylor & Francis Books

Contents

Tables

Acknowledgements

Many individuals and organisations have helped me with this book, and I would like to acknowledge their contributions and assistance. I would mention only a few names but others have also provided valuable help in the course of my writing. I particularly thank Professor, Philip Bell, who steered my attention towards comparing climate change news. Professor Wendy Bacon also contributed to the original study, which is the basis of the book. Thanks also to Dr Catriona Bonfiglioli for her spirited conversation and members of the Climate Justice Research Centre for exposing me to the issue of climate justice. I'd like to especially thank Associate Professor Cherry Russell and Dr Estelle Dryland for their valuable help with editing my writing. I also thank Afroza Shoma for her assistance with data collection from Bangladesh. My sister Sangyukta Das deserves special recognition for her inspiration to get me here. So does my son Shoryu Das-Zaman, who helped me throughout the project, and my partner Akhteruz Zaman, who supported.

1 Comparing climate change news

This book presents an examination of climate change news and views in two vastly different contexts—Australia and Bangladesh. Through an analysis of sources in newspaper articles, it aims to assess the significance of country-specific differences in the knowledge production of climate communication. The comparison is a study in contrasts: Australia is a high-income, industrialised, geographically large country in the southern hemisphere with a low population that is able to claim status as an "honorary Western country" (Curran & Park 2000, p. 3). In contrast, Bangladesh is a poor, agriculture-based, small, low-lying, wet and densely populated post-colonial country in South Asia. Yet, climatically, both are exposed to profound challenges from the effects of climate crises. In reporting on these crises, journalism as a political, institutional and cultural practice has a significant role to play. The universal nature of climate change can erase many differences, but more attention needs to be paid to the diverse ways in which knowledge comes to be made in different places (Hulme 2009; Beck 2010; Eide & Kunelius 2017). Journalism provides information, thus contributing significantly to the construction of public knowledge (Kovach 2006; Patterson 2013). It also contributes to the "processes of public dialogue and societal communion" (Cottle 2000, p. 442). In the process of knowledge production and dialogue, who gets to speak and who does not is a significant determinant of the journalists' capacity to establish authority and assign cultural meaning to realities.

Why contrasting comparison?

The book is comparative, critical and, in a limited way, historical. It assumes that the differences between journalistic practices across the world should not be taken for granted, but empirically explored and tested. It focuses on the news coverage of global and local climate change issues in these two countries. According to many world leaders, climate change is the biggest challenge facing humanity in the 21st century. Although similarities and differences are simultaneously emphasised in comparative studies of journalism, the differences are of particular interest to this book because, while "the similarities are relatively easy to enumerate ... what is really interesting are the differences, however we account for them" (de Burgh 2005, p. 1). De Burgh (2005) further maintains that, while professional practices and organisational format of news media may be similar across the globe, the news content and its format are significantly influenced by the location and cultural contexts of journalism (see also Wasserman 2017). This influence makes country-specific differences in terms of location, culture or history significant considerations in any comparative investigation.

In comparing these two diverse countries, this investigation employs Esser & Hanitzsch's (2012) classification of the comparative journalism research into three categories: actor-based, structural and cultural. The actor- or behaviour-centred approach looks at how professional individuals, groups or institutions make choices in public communication. The structuralist or institutional approach focuses on the broader context, for example, the technological, social, political and economic environment that expedites or restricts communication between actors. The culturalist or interpretative approach considers actors' communication practices as processes of shared meaning between the practitioner and her/his socio-cultural context (Donsbach 2010; Donsbach & Klett 1993; Hanitzsch & Donsbach 2012). Studies of political communication simultaneously apply actor-, cultural- and structure-based concepts to develop a comprehensive understanding of this particular communication process in comparative contexts (Pfetsch & Esser 2012; Tiffen et al. 2013). These three categories are used as a framework in this book to integrate various theoretical approaches to comparative study, and to identify the close interaction of these approaches rather than assume their mutual exclusivity (Esser & Hanitzsch 2012).

Universalist vs culturalist comparison

Critics who have applied these conceptual frameworks in comparisons of national systems of journalism have identified two distinct streams of scholarship: a "universalist" approach, which aims to produce generalisations based on US experiences and assumes that these generalisations are universally applicable; and a "culturalist" approach, which argues that social reality cannot be understood without first exploring contexts that are conditioned by spatial and temporal factors (Hantrais 1999). Inspired by the "universalist" approach, some scholars argue that there are many similarities in journalism practices around the world. Others, however, argue that the differences in professional norms are more significant than the similarities. Hugo de Burgh (2005), for instance, claims that the differences in comparative studies are more interesting and pertinent than the similarities. When it comes to knowledge creation about climate change, Hulme (2010, in Eide & Kunelius 2017) also attaches significance to the "different ways knowledge comes to be made in different places" and expresses concern about the insensitivity of global knowledge to local issues.

Those who favour a "similarity" viewpoint posit that journalists across the spectrum aspire to be independent and autonomous, irrespective of their diverse socio-cultural and institutional differences. These critics further suggest that a crucial way of ensuring independence from other social institutions is through the norm of objectivity. Investigations of objectivity in mononational contexts, such as in Brazil (Herscovitz 2004), Germany (Weischenberg et al. 2012), Indonesia (Hanitzsch 2005), Bangladesh (Ramaprasad & Rahman 2006) and the United States (Weaver et al. 2007), have supported the "similarity" assumption in journalistic practice. Comparative analyses (Deuze 2002; Shoemaker & Cohen 2006; Hanitzsch et al. 2012) provide a broad view of the influences on journalistic practices of professional norms such as newsworthiness and organisational imperatives across different countries. However, scholars in the "culturalist" stream (e.g., Berkowitz et al. 2004; Donsbach & Patterson 2004) remain unconvinced by the similarities identified in those studies. They argue that there are some substantial differences in newsroom practices across the globe, due to the myriad social and political factors that influence news production and, ultimately, shape its content.

Eide & Kunelius (2010) tapped into these myriad factors and found that a "domestication" frame predominated in the coverage of the Copenhagen climate summit (COP15) in 18 countries from both the global North and South. The use of a "domestication" frame has two different consequences. On one hand, it empowers journalists to reach their national publics. At the same time, however, it "restricts the resources, connections, sources of information and possible alliances on which journalists may participate in the construction of a (even momentary) transnational public sphere or space" (p. 41). Similarly, Painter's (2017) extensive analysis of climate change coverage in IPCC news articles in 2013 and 2014, which eventually included five developing and five developed countries, found that the "disaster" theme predominated. These two significant findings in climate communication in the context of global climate events reflect the "universalist" approach, with its associated weaknesses and strengths. One of the advantages of such large-scale studies is the potential to identify "universal laws" (Esser & Hanitzsch 2012) that can be generalised. But, as Livingstone (2003) argues, many phenomena are understood in terms of particular systems (e.g., media systems, education systems). For her, large-scale comparative analysis threatens to undermine "the legitimacy of the nation-state not only for political, economic or cultural purposes but also as a unit of analysis" (Livingstone 2003, p. 480).

Comparative studies have contributed to a growing awareness of journalism across national contexts. However, the selection of the unit of analysis or country suffers from what Curran et al. (2010) term "ideal typification," that is, the selection of historically well-established and similar media systems to understand the commonalities and differences between them. The current study seeks to go beyond the existing rigidity of unit selection to advance an understanding of journalistic practices by comparing dissimilar countries. As some critics (Benson 2010; Blumler & Gurevitch 1975) argue, such divergent comparison is important in that it has the potential to "render the invisible visible" (Hallin & Mancini 2004) in a useful way. The invisible, according to these critics, is the significance of the divergent journalistic practices. To achieve this goal, it is imperative to select dissimilar cases. However, conducting a dissimilar comparison is difficult because such an approach

> can pose challenges to scholars' preconceptions and is liable to be theoretically upsetting ... [But its contribution] is not confined only to testing, validating and revising existing theory. It also has a more creative and innovative role—opening up new avenues.
>
> (Blumler et al. 1992, p. 8)

One of the new avenues for journalism studies is to avoid reductive limitations and pave the way to understanding journalism practices both in their diverse contexts and in a more inclusive manner. To this end, it is crucial to take contrasting countries as the units of analysis because there are significant inadequacies in the existing levels of understanding of journalism practices. For example, contemporary descriptions of journalism in the "developing" world often assume a lack of professional rigor in these less industrialised countries (Shanahan 2011; Romano 2003). This view of professional weakness can be critiqued (Nassanga et al. 2017) as simplistic, because it is a "universalist" outlook that fails to consider adequately the contours of the actual journalism practices in developing countries. As well, it fails to provide any intrinsic reasons for this perceived lack of professional rigor. Such inadequacies can be overcome by simultaneously scrutinising news practices (selection of sources, for example) in both developed and developing countries through a comparative study.

Why does journalistic practice matter in climate change coverage?

One of the many roles that journalism plays in society is that of watchdog or "Fourth Estate," which involves journalists in monitoring political rulers and systems of governance in a society. In their efforts to fulfil this role, journalists negotiate newsworthiness by engaging in interactions with actors and agents from various other social institutions, fostering both adversarial and symbiotic relationships. This book focuses on precisely how they engage in such negotiations in climate change news. Critics have labelled these engagements variously as, for example, "mediated democracy," "mediocracy" and "mediated political realities" (Orren 1986; McNair 2000, as cited in Zelizer 2004, p. 145). These terms emphasise the interdependence between politicians and journalists. News media in many societies deem it propitious to interact with political institutions (Alexander 1981; Schudson 2006). They accordingly become embedded in various countries' political systems, exercising their "Fourth Estate" role, which recognises the informal authority of the press to scrutinise the powerful (Hampton 2010).

This book first considers the political systems in both Australia and Bangladesh and examines the relationship between news media and the state to ascertain the position of news media in relation to other institutions competing for influence in society (Benson 2010). This exercise exposes the complexity and multiplicity of factors involved in the news production process, including political factors, and market or institutional imperatives of media organisations. It also reveals which factors exert the most influence on the news production process. The book explores the significance of the presence of various sources, voices or actors and the absence of others in the news pertaining to climate change in the two contrasting countries. The empirical data comprise newspaper content from Australia and Bangladesh collected during the 2009 Copenhagen and 2015 Paris climate summits. Both these meetings—the 15th and 21st Conferences of the Parties (COP) to the 1992 United Nations Framework Convention on Climate Change (UNFCC), respectively—took place during periods of intense climate debate across the globe over a new climate agreement. The analysis reveals a complex web of interactions between news sources and changing media ecology during these critical moments of global climate negotiations. An important empirical and conceptual contribution of the book is to shed light on how journalists in the global North and South mediate climate change messages. The privileging of certain voices by granting them visibility over others is a key aspect. By measuring visibility in terms of presences and absences, the book explores the extent to which the influences are similar or different in the two countries and how the sources' and journalists' communication power conditions public expressions on climate change. In doing so, the book offers an enhanced understanding of Western and non-Western media systems.

Overall, the book argues that news organisations have used environmental issues, such as climate change, to exercise influence in the public sphere and advance particular political agendas or interests (Mann 2000). Here, journalism is revealed as a key element of political contestation, especially in climate change debates in both Australia and Bangladesh. Media organisations actively seek to gain or retain their power to define the reality of climate change, affirming or challenging other agents' perspectives in the public debate. These engagements make the journalists party to the political contestation over how to define the reality of climate change. In other words, they become "interested participants" as opposed to their widely perceived status of "disinterested observers" (Glasser 1992; Maras 2013). As a backdrop to this exploration, one can draw a parallel between climate change and journalism, because both are global in scope. The issue of climate change affects everyone in the world. The same can be said about journalism, because some form of journalism is practised in various ways everywhere in the world (de Burgh 2005; Josephi 2005, 2013). As well, fundamental and universal values underpin them both; for example, freedom and responsibility for journalism (Christians 2015), and responsibility and solidarity in relation to the issues concerning climate change. Both journalism and climate change pose ethical and political challenges (Laksa 2014; Hackett 2017). For these reasons, the argument for a comparison of contrasting cases to tease out the nuances of journalistic practices can also be extended to climate change communication. In this way, nuances and differences in climate change issues and frames in different places can be exposed and comprehensively understood in their respective social, political and communicative contexts.

Significance

The investigation of climate communication across the North–South divide is especially urgent under the 2015 Paris Agreement. In the agreement, both high-income and low-income countries commit to reducing greenhouse gas emissions. The ambition to achieve "net zero carbon" by 2050 means these commitments will intensify in the decades to come. It is critical in the context of this agreement that we gain a fuller understanding of the dynamics of climate communication in low-emitting, low-income countries, such as Bangladesh, as much as in the high-emitting, high-income countries, such as Australia. An ability to draw out parallels within the expected divergence in the journalistic coverage will directly contribute to the development of effective cross-national strategies to address the mounting challenges of climate change. In focusing on journalism and climate change at two key moments in the global policy debate (2009 and 2015), this book offers a grounded understanding of news practices in contrasting contexts.

Background and contribution to the field

The book makes important empirical and conceptual contributions to the emerging field of studies in climate communication. In recent years, several books on the media representation of climate change have appeared, indicating a growing body of scholarship in this area. Despite this increasing interest, critics acknowledge that it is imperative to conduct more comparative research beyond the confines of the Western print media (Schäfer & Schlichting 2014). As elaborated earlier, very little attention has been paid to the empirical comparison of climate change news coverage across the North-South divide. One important early exception, which focused on the wider context of environmental reporting, is the ground-breaking book *Environmentalism and the Mass Media: The North-South Divide* by James Chapman et al. (1997). This book offered a cross-cultural comparison of the production and consumption of environmental stories in India and the UK. However, it did not specifically focus on news sources. To date, there has been a handful of book-length treatments of the sources of climate change coverage from a comparative perspective (Eide et al. 2010; Boykoff 2011; Hackett et al. 2017). These investigations focused on Northern, high-income countries; none attempted to compare the perspectives with their Southern counterparts. The key issue of journalistic sources is also generally overlooked. Boykoff

(2011), for instance, compared the treatment of sources, but between similar Northern countries (US and UK). While climate change as a globally inter-connected problem exposes large-scale inequalities across the world, critics have emphasised the need for a continued focus on comparative studies. In addition, they have argued for more attention to internal "power and conflict dynamics of social inequalities" (Kunelius et al. 2016, p. 9) within single countries. Some studies have made useful contributions to the understanding of climate change coverage. For example, Kunelius et al. (2016) compared 22 countries, thereby expanding the scope of this scholarship. However, the internal and external factors that influence the production of climate change journalism were beyond the scope of these studies.

This book addresses these gaps in the literature by scrutinising the use of sources and the role of journalistic practices in the politics of climate change. It does so by responding to some empirical questions, such as: Who do jour-nalists use or cite when constituting climate change news issues? How exactly do they quote these sources and represent them in the news? To what extent and in what ways does the journalistic production of climate change repro-duce existing trends in climate change reporting? To what extent does the use of particular voices produce specific types of climate change news? What are the variations in the use of sources in the two periods of this study (2009 and 2015)? The book assumes that the interconnection between increasing envir-onmental degradation and shifting media ecology across the world should not be taken for granted, but explored and tested empirically. The current exam-ination of climate change news thus establishes a "dialogic setting" for understanding heterogeneous and uneven practices of journalism (Wasserman & de Beer 2009). This study of news from contrasting national contexts has the potential to contribute significantly to this field, rectifying the over-reli-ance of climate communication research on a few Western countries for its empirical evidence and theory building (Curran & Park 2000). As noted, such studies are urgently required given the expanded scope of emissions reduction commitments under the 2015 Paris Agreement, which now encompass low-income as well as high-income countries.

Outline of the book

This book establishes the significance of a comparative investigation of the news content on climate change. It discusses the conceptual importance of comparative studies between diverse geographical regions and highlights the need for understanding news media practices across contrasting national and socio-economic contexts. It also recommends conceptual and empirical guide-lines for selecting two diverse countries with interconnected environmental issues, which may be suitable for systematic comparative systematic scrutiny. This chapter has established the significance of a comparative investigation of the news content on climate change in Australia and Bangladesh. As well as discussing the conceptual importance of comparative studies between diverse geographical regions, it establishes the empirical focus of the book, which is on two aspects of the selected news content: who are the dominant sources of climate change news in the two countries, and how are these sources scrutinised or verified? Chapter 2 ("Environmental, Political and Media Systems") delineates the two countries' political and media systems, as well as their environmental policy contexts. It begins with a description of the climate change issues facing these countries, then focuses on the debate surrounding anthropogenic climate events. Brief attention is directed towards the political and media systems of these countries and the relationship between them, which provides the ground for the comparative analysis. It discusses the differences in these countries' environmental contexts as well as broad and distinctive similarities in their political and media systems. In Chapter 3 ("News Sources, Journalism and the Study"), the importance of various sources, such as politicians, experts, government officials and social activists, in the news production process is outlined. The focus is also on the conceptual basis of the cross-national investigation. The beginning of the chapter describes the significance of sources in journalistic production and explains why an examination of the "cross-checking" approach was selected for particular scrutiny. In other words, it discusses how various sources are positioned in the news media's representation of climate change. The chapter also addresses the relative benefits of quantitative and qualitative analyses for examining selected news content and the limitations of these analyses.

The next three chapters present the empirical findings of the study. Chapter 4 ("Climate of Interpretation: Australia and Bangladesh") compares the patterns of editorial opinions pertaining to climate change across the four newspapers. Drawing on Fairclough (2003) and Van Dijk's (1993) discourse analysis and Beck's concept of risk, it also discusses the presence and absence of topics related to climate change in both countries. The chapter interprets the findings through qualitative case analyses of editorial columns. It shows that the news media in Bangladesh tend to portray the country as an adaptive and innovative nation and dispute "Western ecological neo-imperialism" (Giddens 2011). In contrast, Australian news media tackle acute partisanship in climate policy debate, which is heavily influenced by a coal industry that is the third biggest exporter in the world. The comparison between 2009 and 2015 shows the traces of political cosmopolitanism emanating from two discursive positions: economic rationality and ecological vulnerability. Chapter 5 ("Sources of Australian Climate Change News") discusses how Australian journalists drew on various sources to frame the news coverage of climate change in *The Sydney Morning Herald* and *The Australian*. The qualitative and quantitative analyses of the selected content reveal different journalistic strategies employed in the selection and representation of sources. The chapter examines the flow of information between different interest groups, such as politicians, experts, activists, businesses and farmers, to establish how the news media responded to these diverse groups. It reveals a dominance of political sources in the articles from both 2009 and 2015, albeit with a significant presence of business sources in the two Australian newspapers.

The chapter also seeks to ascertain how far the news media can uphold the "critical gaze" (Berkowitz 2009) in the coverage and shows that the statements of political sources are among the least verified or scrutinised. Chapter 6 ("Climate Change News and Sources in Bangladesh") continues to focus on the usage of various sources by journalists. It draws on data from *The Daily Star* and the *Prothom Alo* in 2009 and 2015 to explore how journalists shaped various environmental issues by using different sources in the debate. It identifies variations in the presence of sources between the two periods. Political sources were dominant in the articles from 2009, whereas expert sources were prevalent in those from 2015. As evident in the empirical data, the pattern of source dominance was similar in the two countries in 2009 but diverged in 2015. The statements of political sources were also least verified in these two Bangladeshi newspapers. The analysis of the use of sources elucidates the contestation of power between various agents in defining the reality of climate change as a macro-level environmental threat and sheds light on the complexities resulting from the presence of diverse social elements as news sources.

The concluding chapter discusses the significance of various findings in two parts: a comparison of topics and stances of sources in the four newspapers and a critical discussion of the overall conclusions. The news media in this study, through the selection of sources and cross-checking and interpretation of source statements, bolster their respective organisations' ideological stances by "defining the range of meaningful, consensually unproblematic information that makes sense to readers without violating the journalists' habitual news-values." In particular, the final section recapitulates the key findings and explains their implications for the study of journalism and climate change communication. Through this comparison, the book demonstrates that the climate change debates in the North and South aspire to solve the "new global risk" (Beck 2010) from two distinctively different perspectives, although the communicative modus operandi of some social agents bear similar characteristics.

References

Alexander, J. C. 1981, 'The mass media in systemic, historical and comparative perspective', in E. Katx & T. Szecsko (eds) *Mass media and social change*, Sage, Beverly Hills, CA, pp. 17–51.

Beck, U. 2010, 'Climate for change or how to create a green modernity', *Theory Culture & Society*, vol. 27, no. 2–3, pp. 254–266.

Benson, R. 2010, 'Comparative news media systems: New directions in research', in S. Allan (ed.) *The Routledge companion to news and journalism*, Routledge, London, pp. 614–626.

Berkowitz, D. A. 2009, 'Reporters and their sources', in K. Wahl-Jorgensen & T. Hanitzsch (eds) *The handbook of journalism studies*, Routledge, New York, pp. 102–115.

Berkowitz, D., Limor, Y. & Singer, J. 2004, 'A cross-cultural look at serving the public interest: American and Israeli journalists consider ethical scenarios', *Journalism: Theory, Practice & Criticism*, vol. 5, no. 2, pp. 159–181.

Blumler, J. G. & Gurevitch, M. 1975, 'State of the art of comparative political communication research: Poised for maturity?', in F. Esser & B. Pfetsch (eds) *Comparing political communication: Theories, cases, and challenges*, Cambridge University Press, Cambridge, pp. 325–343.

Blumler, J. G., McLeod, J. H. & Rosengren, K. E. 1992, 'An introduction to comparative communication research', in J. G. Blumler, J. M. McLeod & K. E. Rosengren (eds) *Comparatively speaking: Communication and culture across space and time*, Sage, London, pp. 3–18.

Boykoff, M. T. 2011, *Who speaks for the climate*, Cambridge University Press, Cambridge.

Chapman, G., Kumar, K., Fraser, C. & Gaber, I. 1997, *Environmentalism and the mass media: The North-South divide*, Routledge, London, New York.

Christians, C. 2015, 'North-South dialogues in journalism studies', *African Journalism Studies*, vol. 36, no. 1, pp. 44–50.

Cottle, S. 2000, 'Rethinking news access', *Journalism Studies*, vol. 1, no. 3, pp. 427–448.

Curran, J. & ParkM. J. 2000, *De-Westernizing media studies*, Routledge, London.

Curran, J., Salovaara-Moring, I., Coen, S., & Iyengar, S. 2010, 'Crime, foreigners and hard news: A cross-national comparison of reporting and public perception', *Journalism*, vol. 11, no. 1, pp. 3–19.

de Burgh, H. (ed.) 2005, *Making journalists: Diverse models, global issues*, Routledge, London.

Deuze, M. 2002, 'National news cultures: A comparison of Dutch, German, British, Australian, and U.S. journalists', *Journalism & Mass Communication Quarterly*, vol. 79, no. 1, pp. 134–149.

Donsbach, W. 2010, 'The global journalist: Are professional structures being flattened?' in B. Dobek-Ostrowska, M. Glowacki, K. Jakubowicz & M. Sukosd (eds) *Comparative media systems: European and global perspectives*, Central European University Press, Budapest, New York, pp. 153–170.

Donsbach, W. & Klett, B. 1993, 'Subjective objectivity: How journalists in four countries define a key term of their profession', *Gazette*, vol. 51, pp. 53–83.

Donsbach, W. & Patterson, E. T. 2004, 'Political news journalists: Partisanship, professionalism, and political roles in five countries', in F. Esser & B. Pfetsch (eds) *Comparing political communication: Theories, cases and challenges*, Cambridge University Press, Cambridge, pp. 251–270.

Eide, E. & Kunelius, R. 2010, 'Domesticating global moments', in E. Eide, R. Kunelius & V. Kumpu (eds) *Global climate: Local journalisms*, Projekt Verlag, Bochum, Germany, pp. 9–50.

Eide, E. & Kunelius, R. 2017, 'The problem: Climate change, politics and the media', in R. Kunelius, E. Eide, M. Tegelberg & D. Yagodin, (eds) *Media and global climate knowledge*, Palgrave Macmillan, New York.

Eide, E., Kunelius, R. & Kumpu, V. 2010, *Global climate: Local journalisms, a transnational study of how media make sense of climate summit*, Projekt Verlag, Bochum, Germany.

Esser, F. & Hanitzsch, T. 2012, 'On the why and how of comparative inquiry in communication studies', in F. Esser & T. Hanitzsch (eds) *The handbook of comparative communication research*, Routledge, New York, pp. 3–24.

Fairclough, N. 2003, *Analysing discourse: Textual analysis for social research*, Routledge, London.

Giddens, A. 2011, *The politics of climate change*, Polity, Cambridge.

Glasser, T. L. 1992, 'Objectivity and news bias', in E. D. Cohen (ed.) *Philosophical issues in journalism*, Oxford University Press, New York, pp. 176–183.

Hackett, R. A., Forde, S., Gunster, S. & Foxwell-Norton, K. 2017, *Journalism and climate crisis – Public engagement, media alternatives*, Routledge, London.

Hallin, D. & Mancini, P. 2004, *Comparing media systems: Three models of media and politics*, Cambridge University Press, New York.

Hampton, M. 2010, 'The fourth estate ideal in journalism history', in S. Allan (ed.) *The Routledge companion to news and journalism*, Routledge, London, pp. 3–12.

Hanitzsch, T. 2005, 'Journalists in Indonesia: Educated but timid watchdogs', *Journalism Studies*, vol. 6, no. 4, pp. 493–508.

Hanitzsch, T. & Donsbach, W. 2012, 'Comparing journalism cultures', in F. Esser & T. Hanitzsch (eds) *The handbook of comparative communication research*, Routledge, London, pp. 262–275.

Hantrais, L. 1999, 'Contextualization in cross-national comparative research', *International Journal of Social Research Methodology*, vol. 2, no. 2, pp. 93–108.

Herscovitz, H. G. 2004, 'Brazilian journalists' perceptions of media roles, ethics and foreign influences on Brazilian journalism', *Journalism Studies*, vol. 5, no. 1, pp. 71–86.

Hulme, M. 2009, *Why we disagree about climate change: Understanding controversy, inaction and opportunity*, Cambridge University Press, Cambridge.

Hulme, M. 2010, 'Cosmopolitan climates: Hybridity, foresight and meaning', *Theory, Culture & Society*, vol. 27, no. 2, pp. 267–276.

Josephi, B. 2005, 'Journalism in the global age: Between normative and empirical', *Gazette: The International Journal for Communication Studies*, vol. 67, no. 6, pp. 575–590.

Josephi, B. 2013, 'How much democracy does journalism need?', *Journalism*, vol. 14, no. 4, pp. 474–489.

Kovach, B. 2006, 'Toward a new journalism with verification', *Nieman Reports*, vol. 60, no. 4, p. 39.

Kunelius, R., Nossek, H. & Eide, E. 2016, 'Key journalists and the IPCC AR5: Toward reflexive professionalism?', in R. Kunelius, E. Eide, M. Tegelberg & D. Yagodin (eds) *Media and global climate knowledge: Journalism and the IPCC*, Palgrave Macmillan, New York, pp. 257–280.

Laksa, U. 2014, 'National discussions, global repercussions: ethics in British newspaper coverage of global climate negotiations', *Environmental Communication*, vol. 8, no. 3, pp. 368–387.

Livingstone, S. 2003, 'On the challenges of cross-national comparative media research', *European Journal of Communication*, vol. 18, no. 4, pp. 477–500.

Mann, M. 2000, *The sources of social power*, Cambridge University Press, Cambridge.

Maras, S. 2013, *Objectivity in journalism*, Polity, Cambridge.

McNair, B. 2000, *Journalism and democracy: An evaluation of the political public sphere*, Routledge, London.

Nassanga, G., Eide, E., Hahn, O., Rhaman, M. & Sarwono, B. 2017, 'Climate change and development journalism in the global south', in R. Kunelius, E. Eide, M. Tegelberg & D. Yagodin (eds) *Media and global climate knowledge: Journalism and the IPCC*, Palgrave Macmillan, New York, pp. 213–233.

Orren, G. R. 1986, 'Thinking about the press and government', in M. Linsky (ed.) *Impact: How the press affects federal policymaking*, Norton, New York, pp. 1–20.

Painter, J. 2013, *Climate change in the media – Reporting risk and uncertainty*, I. B. Tauris, London.

Painter, J. 2017, 'Disaster, risk or opportunity? A ten-country comparison of themes in coverage of the IPCC AR5', in R. Kunelius, E. Eide, M. Tegelberg & D. Yagodin (eds) *Media and global climate knowledge: Journalism and the IPCC*, Palgrave Macmillan, New York, pp. 109–128.

Patterson, T. E. 2013, *Informing the news: The need for knowledge-based journalism*, Vintage Books, New York.

Pfetsch, B. & Esser, F. 2012, 'Comparing political communication', in F. Esser & T. Hanitzsch (eds) *The handbook of comparative communication research*, Routledge, New York, pp. 25–47.

Ramaprasad, J. & Rahman, S. 2006, 'Tradition with a twist: A survey of Bangladeshi journalists', *The International Communication Gazette*, vol. 68, no. 2, pp. 148–165.

Romano, A. 2003, *Politics and press in Indonesia: Understanding an evolving political culture*, Routledge, New York.

Schäfer, M. S. & Schlichting, I. 2014, 'Media representations of climate change: A meta-analysis of the research field', *Environmental Communication*, vol. 8, no. 2, pp. 142–160.

Schudson, M. 2006, 'The trouble with experts—And why democracies need them', *Theory and Society*, vol. 35, no. 5–6, pp. 491–506.

Shanahan, M. 2011, 'Time to adapt? Media coverage of climate change in non-industrialised countries', in T. Boyce & J. Lewis (eds) *Climate change and the media*, Peter Lang, Frankfurt, New York, pp. 145–157.

Shoemaker, P. J. & Cohen, A. A. 2006, *News around the world: Content, practitioners and the public*, Routledge, New York.

Tiffen, R., Jones, K. P., Rowe, D., Aalberg, T., Coen, S., Curran, J., Hayashi, K., Iyengar, S., Mazzoleni, G., Papathanassopoulos, S., Rojas, H. & Soroka, S. 2013, 'Sources in the news', *Journalism Studies*, vol. 15, no. 4, pp. 1–19.

van Dijk, T. A. 1993, 'Principles of critical discourse analysis', *Discourse & Society*, vol. 4, no. 2, pp. 249–283.

Wasserman, H. 2017, 'Professionalism and ethics: The need for a global perspective', *Journalism & Communication Monographs*, vol. 19, no. 4, pp. 312–316.

Wasserman, H. & de BeerA. S. 2009, 'Towards de-Westernizing journalism studies', in K. Wahl-Jorgensen & T. Hanitzsch (eds) *The handbook of journalism studies*, Routledge, New York, pp. 428–438.

Weaver, D., BeamR., Brownlee, B., Voakes, P. & Wilhoit, G. C. 2007, *The American journalists in the 21st century: U.S. newspeople at the dawn of a new millennium*, Lawrence Erlbaum, Mahwah, NJ.

WeischenbergS., Scholl, A. & Malik, M. 2012, 'Journalism in Germany in the 21st century', in D. H. Weaver & L. Willnat (eds) *The global journalist in the 21st century*, Routledge, London, pp. 205–219.

Zelizer, B. 2004, *Taking journalism seriously: News and the academy*, Sage, London.

2 Environmental, political and media systems

The debate surrounding climate change, which customarily occurs within the purview of the scientific or economic logic of the environment, has a profound and reciprocal effect on a country's political and media systems. Exploration of these systems is pertinent to understanding the values that underpin the conditions of climate change communication. They not only provide a background for the news content of climate change, but also enable a grounded understanding of the causations behind the production of news. Thus, to understand climate change news in countries worldwide, it is first essential to examine their environmental, political and media systems. What follows is a discussion of these three conditions of climate change news in Australia and Bangladesh.

In the contemporary environment of "increased mediatisation of politics," the relationship between politics and the media has assumed an enhanced significance, i.e., become intense and complex (Hallin & Papathanassopoulos 2002; Schudson 2002). It is undoubtedly beneficial to understand the characteristics of news content and the ways in which news obtains its "internal validity" (Tuchman 1976, p. 97) through the selection of topics and sources and verification of them in the news content. Throughout the world, news emanates from media systems with the aim of influencing democratic political participation and informed debate in the public sphere. In the process, news content invariably engages in political contestation. However, it is difficult to determine which system—political or media—exerts more influence on the final outcomes of contestation. It is not the purpose of this discussion to argue which system exercises the strongest influence over—or poses the strongest challenge to—this contestation. Rather, this brief overview of the environmental, political and media systems of the two countries alluded to above highlights the complex relationships that obtain between the three systems. Exploration of these three systems will facilitate an understanding of how journalists select issues and sources and determine the representation of them in the news coverage of climate change. As earlier stated in the Introduction (Chapter 1), the contexts of climate change in Australia and Bangladesh set the ground for a comparison of the coverage patterns of the topic. Both countries have been labelled "worst-affected" by climate change in their own specific ways. In recent times, Australia has experienced some of its worst periods of drought since colonisation (Potter & McKenzie 2007; Jennings 2012). Bangladesh has been singled out as a potential victim of global warming due to the ever-present danger of rising sea levels (Douglas et al. 2001; Nicholls et al. 1999; World Bank 2018). Keeping this broad categorical similarity in mind, the following discussion outlines the afore-mentioned conditions of climate change news in the two countries.

Climate change in an "El Nino" continent

Australia, the sixth largest country in the world, is a large "El Nino continent" (Grove 2007) located in the southern hemisphere. Due to the vastness of this island continent, it has diverse climatic patterns that are manifested in the tropical nature of the north, the arid environment of the south and a moderately wetter environment in the south-eastern areas in which the major cities are located. Approximately 80 per cent of the land is dominated by arid weather, with annual precipitation of less than 600 millimetres (www.bom. gov.au). As Don Blackmore, former Chief Executive of the Murray–Darling Basin Authority observed: "[T]he landscape [of the Basin] is driven by temperatures, not rainfall" given that only 2.4 per cent of the rain ends up in the basin (Wahlquist 2008, p. 5). His observation is equally applicable to many other parts of this vast continent. Thus, it seems logical to suggest that Australia, with its variable climate and extreme weather events, is vulnerable to climate change.

Much of the debate surrounding climate change in Australia has been based on investigations carried out by the Commonwealth Scientific and Industrial Research Organisation (CSIRO) and the Bureau of Meteorology (BOM). During these debates, the following two major questions have been posed: (1) How and in what ways can Australia navigate climate change? (2) To what degree is the country's average temperature increasing? One recent report stipulated that: "[B]y 2030, Australian annual average temperature is projected to increase by 1.3 degrees Celsius above the climate of 1986–2005" (CSIRO 2015a, p. 6). Rising temperatures accompanied by increased evaporation and subsequent water shortages triggered the "first climate change drought in Australia" (Wahlquist 2008; Saul et al. 2012). In its aftermath, Gross Domestic Product (GDP) declined by 1 per cent. Simultaneously, the gross agricultural product fell by 28.5 per cent, which at the time was the monetary equivalent of 7.36 billion Australian dollars.

Experts have identified recent extreme weather events, e.g., Victoria's "Black Saturday" bushfires in 2009 and the Queensland floods in 2011 as consistent and expected outcomes of climate change (Garnaut 2011). A recent CSIRO (2015b) case study confirmed that the observed changes, for example more frequent hot days, rising sea levels, declining snow depths, intense rainfall events and the increasing acidity of the ocean will continue in the future. Ocean acidity has been declared the cause of another significant indicator of climate change: the coral bleaching of the Great Barrier Reef. Due to rising temperatures and the acidification of the ocean, the monetary loss has been estimated to be two billion dollars per year, particularly in the sustainable fishing and tourism sectors (Wooldridge 2009). Although Australia is geographically vast and economically advanced, its habitat pattern is far from diverse. Some 80 per cent of the populace lives in the coastal areas where rising sea levels pose a direct threat to fixed assets worth approximately 25 billion dollars (Saul et al. 2012, p. 36).

In 2008, the Garnaut Climate Change Review outlined rising temperatures and their impact on the Murray–Darling Basin which was already experiencing little runoff and heavy evaporation. At the time, declining rainfall in the continent threatened to exacerbate the condition of the Basin. In the same year, the CSIRO (2008) predicted that by 2030 the Basin will face lack of availability of surface water. And, changing rainfall patterns will be unpredictable. Due to the scarcity of water, the cost of climate change in the agricultural sector could reach one billion dollars by 2030. Water shortages would likely be accompanied by changes to pasture, increased numbers of pests and diseases, and more extreme events, e.g., fierce bushfires, intensified floods and frequent cyclones. Although Australia is widely viewed as a dry continent, the usage of its water resources is not always prudent. The constant scarcity of water exacerbates environmental problems. For example, the country's river waters are heavily drawn on for agricultural, industrial and domestic purposes. As a result, the protection of "environmental water" is affected and, by extension, the conservation of various ecosystems surrounding the water bodies is hampered. A 2003 audit of land and water resources found significant changes in river conditions in Australia due to intense land usage, increased nutrient and sediment loads in the soil and loss of riparian vegetation (Whittington & Liston 2003). The audited data showed that 30,000 km of river length experienced a high presence of sediment due to widespread clearing of vegetation. The sediment level was 30 cm higher than that of European settlements. The audit also found that more than 85 per cent of the country's rivers were degraded by human activity.

These impacts of climate change provide fertile ground for understanding the tensions observed in the relevant discourses. Despite the evident impacts of climate change, the focus of the debate in Australia is on energy efficiency in the warming world. What it should be upon is the targeted reduction of emissions, a policy Australia committed to as part of the country's Intended Nationally Determined Contribution (INDC). Today, the national discourse is grappling with these two contrasting policies. Policy debate in Australia is consistently underpinned by implicit climate change denial. While on the one hand, primary focus is on energy efficiency both in terms of source (i.e., renewable vs. non-renewable) and price, on the other, attention is on drastically reducing the emissions target committed to via the INDC. Some critics argue that Australia's commitment to the INDC obfuscates the National Energy Guarantee's (NEG) trilemmas of reducing emissions under the Paris Agreement, mitigating increasing electricity prices and ensuring reliable energy supply so that "lights don't go out again" (Byrne 2017). They fear that the NEG would make each of the three trilemmas even worse.

The split in the focus of the national debate defies all the evidence presented regarding the Australian environment. In addition, it reinforces the climate sceptics' claim regarding the credibility of climate science (Garnaut 2011). The relentless doubt about climate change persists, both in common parlance and media discourses. For example, in his controversial argument against the notion of anthropogenic climate change, Ian Plimer (2009) claims that volcanic eruptions release more carbon dioxide (CO_2) into the atmosphere than human activities. He further argues that current scientific investigation of climate change relies on dubious computer modelling; thus, the findings are extremely under-representative in terms of a broader timeline (Gocher 2010). Despite comprehensive response to these issues from climate scientists and the former Department of Climate Change (2011), sceptics like Plimer, and sympathetic organisations including *The Australian*, the Institute of Public Affairs (IPA) and the Australian Environmental Foundation (AEF) persist with their anti-climate-change campaigns (Media Watch 2012). News Limited's flagship publication *The Australian* continues to challenge scientific evidence associated with climate change, despite simultaneously focusing upon issues including damage to the Great Barrier Reef, the Victorian bushfires of 2009 and the Murray–Darling Basin crisis (Media Watch 2010; Nash et al. 2009). Another manifestation of climate scepticism in Australia has been the recent formation of a "ginger group" within the Turnbull government, popularly known as the "Monash Forum." This informal government faction, which includes former Prime Minister Tony Abbott among its members, agitates for the government to review its current Energy Policy, by extension paving the way for the construction of a new taxpayer-funded coal-fired power station. The forum insists that coal-fired generators will bring energy prices down in line with the government's NEG plan. However, many commentators who dispute this claim argue that the evidence suggests the contrary (Parkinson 2018; Hudson 2018). The resurgence of the coal-fired power station debate may be impeding any commitment made by Australia at the Paris Conference of Parties in 2015 vis-à-vis reducing emissions.

Climate change in "nature's laboratory"

According to a panel of international experts,

> climate change poses significant risks for Bangladesh, yet the core elements of its vulnerability are primarily contextual. Between 30–70% of the country is normally flooded each year. The huge sediment loads brought by three Himalayan rivers, coupled with negligible flow gradient, add to drainage congestion problems and exacerbate the extent of flooding. The societal exposure to such risks is further enhanced by Bangladesh's very high population and population density.
>
> (OECD 2003, p. 6)

International experts claim that the exponential growth of Bangladesh's population is significantly impacting the country's environment. While local experts do not discount the impact of population growth, they place more emphasis on the impact of climate change which, they claim, further aggravates a situation already affected by a range of natural disasters. Professor Ainun Nishat, a local expert working with the International Union of Conservation of Nature (IUCN), warns that Bangladesh is facing imminent danger from encroaching seas, fiercer and more frequent cyclones, and storms and floods, all considered due to the relentless rise in global greenhouse gas emissions in the atmosphere. Professor Nishat, who alludes to Bangladesh as "nature's laboratory of disasters," explicates his choice of nomenclature as follows: "We don't have volcanoes. But any other natural disaster you think of, we have it" (Inman 2009, p. 18). Local experts are also critical of inadequate government response to many ongoing environmental problems, e.g., shrimp farming, the illegal felling of trees in forests and the indiscriminate poaching of wildlife. This criticism is consistent with the findings of some studies, an outcome that will be discussed in later sections of the book.

As predicted by the Intergovernmental Panel on Climate Change (IPCC 2007) and reinforced by the 2014 World Bank Climate Change Report titled "Turn Down the Heat: Confronting the New Climate Normal," the effects of a rise of 4°C (7.2°F) in temperature by the end of this century would cause a sea level rise of three feet in Bangladesh. Such a rise has the capacity to jeopardise the habitats of 24 million people in the country's coastal regions (World Bank 2014). In line with numerous previous reports, the World Bank report issued the following stern warning: "[I]f warming continues unabated, irreversible changes [on] a large scale could be triggered" in the form of rising heat extremes, changing precipitation and melting glaciers across different areas of the world. This would pose a critical risk to the development of less resourceful regions. As a result, the world's poorest nations would suffer the worst consequences due to rising sea levels, shifting water resources, declining crop production (and consequent severe food shortages), and hunger and poverty (Germanwatch 2012, 2018; OECD 2003). According to the latest Global Climate Risk Index 2018 produced by the think-tank Germanwatch, Bangladesh continued to rate sixth among the world's most affected countries during the period 1997–2016. With two-thirds of its land less than five metres above sea level, Bangladesh is immensely vulnerable to climate change. Much climate-related research has exposed the vulnerability of Bangladesh to warmer weather and increased precipitation in the Ganges–Brahmaputra–Meghna (GBM) Basin. If the predicted rise in global temperature from 1.4°C to 2°C eventuates by 2050, precipitation could increase from 2 to 7 per cent. Concomitant with a single degree temperature rise, rainfall would increase by 5 per cent in the GBM Basin area, and the intensity of flooding would escalate simultaneously. Should this occur, 20 per cent more land area in Bangladesh risks being flooded (Dasgupta et al. 2011). The IPCC's fourth assessment (2007) cautioned that while the frequency of tropical cyclones could increase by 5–10 per cent, the locations of cyclones would remain the same. The consequent intensity of cyclones could result in strong wind gusts and greater exposure of coastal land to storm surges. The predicted extent of sea level rise has been an important topic of discussion in Bangladesh. Different models have variously predicted sea level rise in this low delta land. For example, the Bangladesh government calculated a rise of 30–100 centimetres, and a rise of "14, 32 and 88 cm for the years 2030, 2050 and 2100 respectively" (Karim & Mimura 2008, p. 493; also mentioned in IPCC 2007). To date, there is currently no precise indication of a specific rise in sea level for Bangladesh. Some experts consider such a prediction difficult due to the active nature of the Ganges–Brahmaputra delta, with its dynamic morphology. In addition, local investigation of the river systems revealed that Bangladesh would gain 640 square kilometres of land through accumulation of sedimentation from the river systems (Inman 2009). Although this prediction potentially challenges the IPCC findings, it is not clear how the limited height of these lands could combat rising sea levels.

The idea of Bangladesh's "contextual" vulnerability has attracted the attention of critics in recent times. Rashid and Paul (2014), who contest the notion of "contextual" factors, argue that the factors are no longer confined to standard textbook descriptions of "atmospheric" or "hydrological" hazards. Rather,

> In the context of climate change, these hazards and disasters are all driven by atmospheric processes although the nature of the hazards differs significantly from one another. Some of them are directly atmospheric processes, such as hurricanes (tropical cyclones) and tornadoes, while others, such as storm surges, river floods and flash floods are the outcomes of atmospheric processes.
>
> (Rashid & Paul 2014, p. 21)

As part of the delta land, Bangladesh has been continuously exposed to both severe and less severe climatic disasters. Major climatic disasters, including tropical cyclones, storm surges, river floods and droughts are often associated with significant economic loss and displacement. Other less severe climatic hazards include tornadoes, pre-monsoon nor-wester windstorms and flash floods. The latter, which are sometimes the direct consequences of riverbank erosion, are considered geological processes. Such erosions are "directly related to climate-driven hydrological processes, especially changes in the stream-flow regimes" (2014, p. 21). The irony for Bangladesh is that its high vulnerability does not match the low level of emissions this country releases into the atmosphere. Its low emissions can be better contextualised when a comparison is drawn with Australia, a high-emitting country. According to the World Bank's DataBank (2018), the annual per capita carbon dioxide emissions figure for Bangladesh was only 0.5 tonne in 2014, while the figure for Australia was 15.4 tonnes. Despite being a very low-emitting country with a high level of vulnerability, Bangladesh does not appear to be "just sitting on its hands" by attributing the responsibility for climate change to the developed countries. In 2009, the country developed a plan titled "Bangladesh Climate Change Strategy and Action Plan" (BCCSAP), the idea being to share its climate adaptation knowledge with similar countries through South–South exchanges. Through this knowledge-sharing process, local experts including Dr Saleemul Huq (2018) are now keen to spur the developed worlds into action. The developed countries' focus should be upon drastic reduction of emissions in accordance with the respective INDCs' commitments at the Paris Conference of Parties in 2015. In its INDC, Bangladesh has committed to a 20 per cent reduction of greenhouse gas emissions by 2030, 5 per cent unconditional and 15 per cent conditional. The country's ambition is to become a climate resilient middle-income country by 2021. But, despite its heavy emphasis on carbon neutral power plants in the BCCSAP, the government's recent plan to build a coal-fired power plant at Rampal near the Sundarbans mangrove forests raises serious questions regarding its genuine intentions vis-à-vis its emissions reductions plan. The Sundarbans constitute approximately 50 per cent of the reserved forest of the country and are a UNESCO heritage-listed site (Gutierrez 2016). In sum, both Australia and Bangladesh are significantly vulnerable to sea level rise due to many factors including the concentrated habitat pattern in Australia's coastal regions, and the geographic location of Bangladesh in the low-lying delta region (see Diamond 2005).

Australia's political system

Australia's system of government is enshrined in liberal democratic principles underpinned by individualism, religious tolerance and freedom of speech. The country's government institutions are similar in practice to the British and North American models, albeit with their own distinctive features, e.g., a strictly enforced compulsory voting system that is viewed by some as a "democratic innovation" (Jones & Pusey 2010). There is some suggestion of introducing this system elsewhere, particularly the in UK where the voting turnout in elections has dwindled over the years. The Commonwealth of Australia was constituted in 1901 when six former British colonies formed a union after separation from the British Empire. Australia retained part of the imperial tradition in the form of a constitutional monarchy. Currently, the Queen of the United Kingdom remains Australia's head of state, although the monarch's power is limited by the Australian constitution which laid the framework for the Australian system of government. The country is divided into three tiers of governance: local, state and federal. Through the formation of state governments, former colonies retained their power to legislate in certain jurisdictions; education and transport, for example. Both the federal and state governments are divided into three branches: legislative, executive and judicial. The legislative and judicial branches function independently of each other. The legislative has two wings: the House of Representatives (Lower House) and the Senate

(Upper House, also known as the "house of review"). Having two wings ensures that legislation does not discriminate against any one specific state. While the Australian governance system is considered one of the most effective among the Western democracies, some critics are not convinced about its democratic innovation principles. They argue that the scope of democratic practice in the Australian system is very limited (Kuhn 1998). The only occasion on which people can freely exercise their democratic rights is every three years at their local polling stations. The remainder of the time they are at the "receiving end" of the political decision-making processes over which ordinary citizens have limited or no control, although they impact on their lives in many ways (Kuhn 1998, p. 370).

So, who influences the decision-making processes in Australia? This question does not invite a clear-cut answer. Different critics have different views, often depending on the socio-economic perspectives they pursue. The political economy perspective argues that policy formulation is an outcome of bargaining, mostly between political and commercial elites. The critical studies perspective posits that different "public interest advocacy groups" (e.g., Australian Consumers Association, Youth Media Group, Media Entertainment Arts and Alliance) play crucial roles in public policy formulation. This is particularly evident in the formulation of media policies (Gruen & Grattan 1998). Others (Marsh 1995; McKnight 2005; McEachern 1991) argue that issues related to economic matters are often heavily influenced—and at times determined—by a coterie of business managers who are influenced by contemporary "market values." The latter posit that all economic and social issues can be handled with "individualism, competition and [a] free market" (McKnight 2005, p. 18). However, yet more others see the influence of Australian politicians' "business mates" as significant albeit not always apparent. In order "to press their claims in the political arena, business has built a wide variety of organisations which seek to speak on behalf of all or part of business" (McEachern 1991, p. 135). This has been evident in the media's coverage of climate change (for further discussion, see Chapter 5), which consistently demonstrated the power of business lobbies in Canberra to dissuade the government from observing the fundamental principles of climate change policy in 2009 (Maddison & Dennis 2009). Some critics argue that business lobby groups are undermining Australia's political culture which has successfully developed a two-party system (Marsh 1995, 2010). The Australian Labor Party emerged as a mass political movement in 1891. Later, the 1909 merger of the Protectionists and Free Traders led to the formation of the modern Liberal Party by Sir Robert Menzies in 1946. This two-party system has encouraged duopolies and, at times, quasi monopolies (James 2010; see also Curran et al. 2010) in sectors including retail (Woolworths vs. Coles supermarkets), telecommunications (Telstra vs. Optus), banks (the group of four) and print media (News Limited vs. Fairfax). However, a general connection between the two-party political system and duopoly in the media system is not adequately substantiated because in other countries with similar two-party systems (e.g., the US and UK), one finds significant diversity in media and other sectors. As in many other countries, the ideological tension between the two main political parties (Labor and Liberal) is marked because both are predominantly driven by "market values." This ideological indifference is further evidenced by the fact that both parties try to "capture the Middle Ground" (Jaensch et al. 2005); that is, try to win the right to govern by convincing and persuading constituencies of the legitimacy and benefits of their respective policies.

The policy-making capacity of the major political parties puts them in a formidable position when it comes to formulating media and climate policies in Australia. Within this two-party political system, both the Labor and Liberal parties mould their media policies in accordance with their close interaction with influential media proprietors, e.g., the late Kerry Packer and former Australian citizen Rupert Murdoch (Griffin-Foley 2003; Masters 2006, quoted in Cunningham & Flew 2010, p. 32). Similarly, the Labor-Coalition duopoly enjoys a close relationship with the resources sector (e.g., coal mining and energy) as these big businesses and political parties mutually endorse each other. As a consequence, none of these major political parties in Australia is ready or willing to deliver the required emissions reduction to play the country's role in keeping the global temperature rise below 2°C (Holmes 2016).

Political system in Bangladesh

Bangladesh achieved independence from Pakistan in 1971 after a bloody liberation war. Since then, the country has undergone a series of political upheavals. The toppling of its first elected government in a military coup in 1975 ushered in a decade and a half of military rule. In fact, some have argued that Bangladesh has struggled twice for its freedom, first to obtain independence from Pakistan in 1971, and then its move from a military junta to installing electoral governance in the early 1990s. In 1990, mass and student uprisings against the military rulers paved the way for the latter's exit and the re-establishment of democracy. Since then, Bangladesh has had six national elections to its unicameral legislative council, the 350-member national parliament (the Jatiya Sangshad). However, in Bangladesh, the institutionalisation of democratic practices in politics remains tenuous (Khan 2003). Power is still heavily concentrated in the upper echelons of government, a consequence that has engendered widespread corruption across government establishments. By extension, corruption has affected the country's entire population. The mainstream political parties' lack of democratic culture has permeated other areas of social life, resulting in intolerance and a range of negative effects. As Khan writes:

> Both the supreme leaders of the two major political parties, i.e., the Awami League and [the] Bangladesh Nationalist Party, were handed top leadership positions for reasons of heredity and kinship. This permanent nature of supreme leadership thwarted internal democracy in the political parties.
>
> (Khan 2003, p. 391)

Lack of democratic practice has resulted in a wide range of problems, e.g., the politicisation of public bureaucracy, lack of accountability and transparency, and inefficiency of public offices (Zafarullah & Siddiquee 2001). My use of the term "politicisation" here refers to interventions by politicians in various functions of public bureaucracy and other areas including educational institutions and businesses. Politicisation becomes particularly illuminated and criticised when public service officials exchange political loyalty for personal favours, e.g., promotions and lucrative postings. Within this process, public bureaucrats become enthusiastic supporters of party policies. But, this situation impacts negatively on their professional ability to provide efficient policy support to the government. Due to the unchallenged "supreme leadership" convention in party politics, the country's parliamentary system has virtually transformed into a "Prime Ministerial system" whereby policy decisions are frequently confined to the Prime Minister's office. The fact that a Prime Minister enjoys the unconditional and total support of her/his party's legislative council members has given rise to a "highly personalized and centralized style of governance with a strong sense of partisanship" (Kochanek 2000). In theory, while a Westminster-style elected parliament has replaced the previous presidential system, in practice the parliament has become a platform for passing legislation initiated by the government and subject to very little deliberation and scrutiny. Various groups in society, e.g., opposition political parties, international donors for development, and civil society organisations have attempted to scrutinise and influence the legislative process. Many have expressed their concern regarding ongoing crises including widespread corruption and mismanagement in different sectors. Governments in Bangladesh frequently find themselves at odds with civil society organisations, particularly those that are vocal critics of various public policies. Successive governments have accused these organisations of exercising prejudice against the public sector. Their criticism arises from the fact that many operate as "not for profit" non-governmental organisations (NGOs). For this reason, they are accountable only to their overseas financial backers, not to the government. Questions surround the integrity of NGOs: some operate commercial ventures while at the same time enjoying non-profit status. For these reasons, the relationships between the government and civil society organisations becomes tense, particularly when the government accuses the advocacy programmes run by some NGOs of intervening in the country's internal affairs, challenging the legitimacy of the state (Parnini 2006). The accountability and legitimacy of the government's activities are constantly under scrutiny. As Khan (2003) argues, governance in Bangladesh is based upon secrecy, and a fundamental lack of accountability, both of which are endemic in all sectors including the governance of media. It may well be that the root of this secrecy and lack of accountability in the public sector lies in an old draconian law known as the Official Secrecy Act 1923, and its off-shoot provision in the Bangladesh Service Rules. Reminiscent of British rule in India, this law prohibits the release and receipt of any government information without official authorisation.

Despite sharing an historical past as colonies of the British Empire, and despite being current members of the Commonwealth, the political systems of Australia and Bangladesh evince a wide range of differences dating back to the inception of the two countries. Whereas Australia was separated from the British Empire in 1901 through political negotiations, Bangladesh (formerly East Bengal) severed its links with Britain in 1947 after a long and often turbulent independence struggle. The sub-continent was partitioned into India and Pakistan, and East Bengal became East Pakistan. Later, in 1971, the country gained independence from Pakistan following a bloody war of liberation. Over time, however, conditions in Bangladesh have deteriorated due to frequent natural disasters, a burgeoning population and poor governance. Australia, on the other hand, has enjoyed a stable period of economic growth as well as healthy political practice, making it one of the longest-serving continuous democracies in the Western political system. Australia's political system has remained consistent regarding the primacy of the private sector. This has helped its economy to maintain a competitive edge in the global market. As well, it has ensured a strong relationship between business and politics. Examination of Bangladesh's economy reveals an overall dominance of politics in society. This is visible in the politicians' ability to intervene in private sector business affairs, a practice suggesting that a certain level of reciprocal yet irregular relationship exists between business and politics. These differences in the two countries' political systems have obvious implications for their respective media practices. However, close examination reveals little obvious manifestation of difference in the media systems of the two countries. Their private and public media organisations coexist with each other, albeit with varying degrees of audience access, overall influence and regulatory regimes.

Media systems

The following section considers the different media systems across the world, and identifies some similarities and differences among them, particularly in the sphere of professional media practices and norms. Discussion of the media systems will facilitate an understanding of the professional practices relevant to the use of news sources in Australia and Bangladesh. Scholarly insights, e.g., "all states and media systems are authoritarian; it just depends upon who is the authority—political power or public sanction" (Merrill 2004, pp. 14–17), will help position the subsequent discussion about media systems in the two countries in a broader context.

For the purposes of this discussion, I have drawn from a series of analyses of media systems including those of Josephi (2005) and de Beer & Merrill (2004). This corpus of work can be divided into three streams. Although they are not chronological, they are interspersed with each other at differing levels of analysis. First, the classical stream, which commenced in the mid-1950s, was characterised by Siebert, Peterson & Schramm's (1956) seminal work titled *Four Theories of the Press*. The authors viewed the news media's role in society from a political perspective, then divided the world of journalism into the following four categories: the authoritarian model (early-modern England, contemporary Iran, Paraguay, Nigeria); the libertarian model (United States, Japan, Germany); the communist model (China, Cuba, North Korea, Russia); and the social responsibility model which emerged from the United States. But, to date, no country or region fits into this social responsibility model. From the mid-1950s to the early-1990s, several generations of scholars and journalists were heavily influenced by the philosophical assumptions of the "Four Theories" model. They subsequently conducted a range of investigations and produced numerous international textbooks (McQuail 1994; Wright 1976), their aim being to provide a "panoramic view" of journalism around the world (Merrill 1983, p. xi). One of the strong criticisms of this landmark Four Theories study—and subsequent investigations—is that they had attempted to universalise the experiences of the United States and Great Britain as models representative of the rest of the world. This approach viewed the supremacy of Western capitalist democracy as a political ideology. In subsequent investigations undertaken during the 1990s, i.e., the period that saw the second stream, scholars challenged the Four Theories model criticising it as monolithic and demonstrative of a "lack of knowledge about other media systems" (Curran & Park 2000, p. 4). In their seminal work titled *De-Westernizing Media Studies*, James Curran and Myung-Jin Park observe that the normative values of Western media practices, e.g., liberty, equality and solidarity, seemed to have universal appeal. However, the dynamics of the relationship between media, state and the public varied across different regions and countries (Merrill 2004). In this stream, focus was upon the media, society and political systems. In other words, the main question driving these studies was: what influences and controls the media? Scholars advocating this stream adopted an interpretative qualitative methodology for the purposes of their investigations.

Hallin & Mancini (2004) propose three models of media systems: (1) the Polarised Pluralists model practised in France, Italy and Spain; (2) the Democratic Corporatist model found in Finland, Switzerland and Germany; and (3) the Liberal model, which they saw as prevailing across Ireland, Britain and North America. These three models were considered yet another initiative demonstrating that the "Anglo-American model is not the one that fits the rest of the world" (Josephi 2005, p. 580). Here, the Polarised Pluralist model stepped away from the conventional way of looking at the practice of journalism, creating scope for developing countries to explain their journalism beyond the existing dominant frame of development journalism, which emphasised the key role of journalism in fostering national development (Xiaoge 2009). Critics raised questions about the applicability of these models to actual situations; as well, they questioned their implications for the existing Western model of journalism across the world. It was from this level of analysis that critics (e.g., Josephi 2005; Wasserman & de Beer 2009; Hallin &

Mancini 2012; Curran et al. 2010) shifted the focus of analysis from the political purpose of journalism to more specific practices of news work. In other words, the ways in which different tasks in the production of news are performed today have been the focus of the above studies. These studies can be identified as the third stream of investigation of media systems. The above three streams of research provide the background for the comparisons made in the empirical sections of this book between newspapers from the first and third worlds. This background puts into context the following brief outlines of the media systems in Australia and Bangladesh.

Media system: Australia

Australia, which has existed as a modern state for more than two centuries, has experienced a strong presence of newspapers due to the advent of mechanical printing in the mid-nineteenth century (Lawe-Davies & Le Brocque 2006). From the 1920s on, the gradual introduction of radio and television broadcasting in the public, private and community sectors has seen the news media develop into a vibrant market. Despite the diversity of its various media, the media system in Australia has been characterised by the most concentrated ownership of commercial media in the developed world (Tiffen 2010; Josephi & Richards 2012; Independent Inquiry into the Media and Media Regulation 2012). Concentration is particularly evident in the sphere of newspaper ownership, wherein Australia is ranked highly compared to other like nations (Tiffen & Gittins 2009; Harding-Smith 2011; Independent Inquiry into the Media and Media Regulation 2012). As well as the press, concentration of ownership also has a place in free-to-air broadcasting. Although Australia has developed a "dual system" of broadcasting ownership—combining a US-style commercial operation with a UK-style public broadcasting system—ownership in the commercial sector is highly concentrated, limited to a handful of operators only. The introduction of this "dual system" has not been without contestation. While some labelled it "the best of both worlds," adding that it contributed to an informed citizenry essential for the success of Australia's compulsory voting system (Jones & Pusey 2010), Josephi & Richards (2012) were openly critical of media ownership concentration in Australia: "One of the side effects of this situation has been intermittent but ongoing tension between public and private media, especially since the establishment of the Australian Broadcasting Commission (later Corporation) in 1932, and the Special Broadcasting Service in 1978" (p. 115). Any attempt to understand the origins of such tension would benefit by trying to determine the position of Australian media structure in the previously discussed media system models.

Hallin and Mancini (2004, p. 7) see Australia's media system as a Liberal model characterised by a democratic political system and the rise of highly-professional and information-based journalistic practices. Jones and Pusey (2010), who contest this categorisation, provide several reasons for arguing against the Australian media system fitting into the North Atlantic Liberal model. Their reasons include the relatively "late professionalisation of journalism" (Henningham 1996), and the low educational qualifications of journalists in Australia compared to levels in other advanced democratic societies (Weaver 2005). Jones and Pusey (2010) add that not only the questionable education levels and professionalism of some journalists, but also certain government policies have significantly impeded the development of the Liberal model. One example is the convention of appointing a Chairman to the Board of Directors of the Australian Broadcasting Corporation (ABC). Often, the government of the day chooses a Chairman according to its political preference. Another example is the controversial "Murdoch Amendment" introduced by Malcolm Fraser's Liberal government "in the late 1970s to facilitate [Rupert] Murdoch's concentration of television ownership" (Chadwick et al. 1995, p. 67). In response, media owners were strongly inclined to intervene in the political world (Chadwick 1996; Griffin-Foley 2003). Interventions of this type confirmed many politicians' observations that media owners had the capacity to influence election results. According to Jones and Pusey (2010), these policy conventions go against the Liberal model of the media system, according to which news organisations should enjoy constitutional guarantee of independence from the government of the day. In answer to their critics (Chadwick 1995; Griffin-Foley 2003), Jones and Pusey (2010) argue that while the Australian media system might not fit into the North Atlantic Liberal model, it could fit into what they term the "Mediterranean Polarised Pluralist Model," a model strongly distinguished by political parallelism. This means that in effect the news media provide unquestioned support to certain political parties, in the process forfeiting (or heavily undermining) their own professional norms. In this respect, the findings of some recent investigations (Bacon 2011; Independent Inquiry into the Media and Media Regulation 2012) are highly relevant to this discussion. The fact that investigations provide a diverse picture of media practices makes it difficult to neatly categorise Australia's media system into one model or the other. News media practices are generally identified as partisan in these studies. For example, Bacon (2011), who examined the news coverage of the Australian government's climate policy, found that News Limited publications campaigned against the climate change policy instead of reporting on it. Fairfax publications, on the other hand, provided a far more balanced coverage of the topic. So, if one takes into consideration News Limited publications, the Australian media system appears to reflect political "parallelism". But, such is not the case with Fairfax and the ABC. The latter is generally considered a trusted source of information by the Australian public (Independent Inquiry into the Media and Media Regulation 2012).

The principle of partisanship in the Polarised Pluralists model becomes more complicated when the perceptions of Australian journalists vis-à-vis their profession are considered. Journalists generally agree that the professional norms to which they adhere give credibility to their profession and institutions (Das 2007; Josephi & Richards 2012). Yet, public confidence in journalism and the media institutions, particularly the commercial institutions, has reached an all-time low. At the same time, the public broadcasting system, especially the ABC, continues to enjoy a high degree of public trust as a reliable source of news and current affairs (Independent Inquiry into the Media and Media Regulation 2012). This somewhat complicated scenario inhibits one's ability to draw a simple conclusion about the professional standing of a country's journalism or media organisations.

Media system: Bangladesh

The media system in Bangladesh is diverse, particularly in terms of ownership; but, the causes of this diversity of both the print and broadcasting media have to date failed to attract scholarly attention. Only recently has discussion surrounding the country's media systems, particularly its broadcast media, started to emerge. This may have been prompted by the following factors: (1) there has been unprecedented expansion of the television industry in Bangladesh since 1997 when the first television channel in the private sector was introduced. By 2009, the number of television stations in Bangladesh reached 19, realising a Gross National Product (GNP) of only $440 per capita (Rahman 2009); by 2015, the number of television stations crossed 40. (2) This rapid expansion had an adverse effect on the practice of journalism. The production of news in Bangladesh has always been at the centre of controversy. All the political parties, along with many civil society organisations, accuse the news media of being biased towards specific political parties (Chowdhury 2003).

By contrast, news media in other non-Western societies have been deemed "effective development change agents." Some critics consider this to be the principal characteristic of Eastern patterns of journalism (Gunaratne 2007; Chowdhury 2003). However, concomitant with the market liberalisation of the broadcast industry, the role of news media has shifted from "development change agent" to one of informing and entertaining audiences. That notwithstanding, the notion of "change agent" is still alive and active in the country's public service broadcasting network, i.e., Bangladesh Television (BTV). From the time of its inception in 1964 until now, BTV's free-to-air terrestrial channel has enjoyed a wide audience base across the country. Its popularity is due in the main to its accessibility in the remote rural regions. The emergent satellite broadcast channels cannot reach these areas due to technical inaccessibility. As well, rural audiences generally cannot afford to subscribe to expensive private channels. At this point, one could argue that the media system in Bangladesh demonstrates dual characteristics: a US-style competitive deregulated private-sector television industry, and British-style taxpayer-supported public service broadcasting.

However, unlike in Australia and the UK, where public service broadcasters are the exemplars of best practice in journalism, BTV has been playing an extremely partisan role, consistently acting as the mouthpiece of the ruling government of the day (Roy 2006). In Bangladesh, the broadcaster attracts severe criticism from many quarters for this role: its professional integrity as a news provider has always been subject to question. However, there is an interesting twist in the operation of this public broadcaster. It sells broadcast time to commercial advertisers, despite being treated as a public agency under the Ministry of Information. As well, its total funding comes from the government's revenue budget. BTV's rate of advertising is much higher than that of other private channels that supposedly enjoy much more professional freedom and standing as news providers. BTV charges up to 75,000 taka (US$1,088) a minute for advertising (Roy 2006), although in recent times the number of TV viewers has been decreasing across various local channels (Irani 2017). This high rate for advertising time demonstrates a contradiction. While on the one hand, despite its public service status and partisan nature, this broadcaster is commercially more viable than the privately owned commercial broadcasters; on the other, the high rate charged for advertising does not necessarily translate into high power for the station. In fact, the editorial policy of BTV news remains subservient to the government of the day. According to the capitalist notion of the press (Curran 2002), economic emancipation is critical for the independence of news media from state control, and it is largely due to advertising revenue. However, this notion of commercial power enhancing the editorial power of a news organisations clearly does not help to explain the case of Bangladesh TV.

Bangladesh experienced satellite broadcasting for the first time in 1992, in the form of transmission from BBC World and CNN International. As of 2017, the country has 43 television stations, including terrestrial and satellite services. This indicates that television services constitute an influential and lucrative area of investment for private ventures. However, competition for the audience share is fierce among these stations, which has negatively affected the channels' news services. News, one of the most popular television genres in Bangladesh, enjoys approximately 66 per cent of audience share compared to other programmes such as drama, cinema and talk shows (AC Nielson 2006). However, due to steep competition, the role of news services has gradually undergone a series of shifts from orientation to "socially responsible" to "market oriented" journalism in this emerging media climate, a phenomenon experienced by many industrially advanced countries (Iyengar 1991; Curran 2002; Hallin 1996). Market-oriented journalism is characterised by personalised and de-contextualised news which focuses more on events and individuals than on the broad underlying causes of various social issues and developments.

Studies of ownership of the television industry in Bangladesh have revealed that the ownership is not confined to "extremely rich people" with strong political connections; it has extended to politicians from the country's two main political parties, the Bangladesh Nationalist Party and the Bangladesh Awami League (Rahman 2007; Roy 2006). From this, it may be inferred that Bangladesh's media system broadly fits into Hallin and Mancini's (2004,

2012) Polarized Pluralist model, which entails partisanship as a fundamental element of the functioning of news organisations (McCargo 2012). Other aspects of this model include strong state intervention, a tendency towards the instrumentalisation of the media by political and economic elites, a weak journalistic professional culture and low newspaper circulation numbers. These aspects are to some degree evident in Bangladesh's news media, albeit with some exceptions and variations. For example, it would be inappropriate to accuse the different media in Bangladesh of having a weak journalistic professional culture. The country's print media, which attract 20 per cent of the total media audience (Chowdhury 2003), have a strong professional tradition. Most of the print news organisations achieve this by not supporting blindly either of the two mainstream political parties. Undoubtedly the different newspapers are influential players in the field of news media in Bangladesh. However, this is not to suggest that all print media operate free from party influences; there are, indeed, some prominent partisan print publications. What I am emphasising here is that widely circulated, high-profile newspapers including *Prothom Alo, Jankantha, Jugantor, Ittefaq, The Daily Star* and *The Daily Independent* have the capacity to play roles beyond the prevailing partisan practices, i.e., to set the agendas for other media organisations.

The role of journalism in this country has been complex, so much so that usage of the polarising terms "neutral" and "participant" (Hanitzsch 2005) seems barely adequate to describe it. Historically speaking, newspapers in Bangladesh have been intricately connected to various prolonged political movements, e.g., gaining independence from colonial British rule and post-colonial Pakistani rule in 1947 and 1971 respectively, and establishing democracy by overthrowing a military dictatorship in 1990. Many journalists, particularly those strongly committed to "social change," are ready to move beyond the professional mindset of the distant observer and neutral reporter to intervene in any situation that requires action. However, distancing themselves from the immediate political turmoil and upholding professional standards of accuracy, fairness and balance does not weaken their commitment to the liberal ideals of objective journalism (Anam 2007; Ramaprasad & Rahman 2006; Rahman 2005). The print media's "middle path" position underpinning journalism's role in social change, together with state-run Bangladesh TV's partisan editorial policy and the highly commercial potential of its products, presents a complex picture of the media system in Bangladesh. This rules out categorising this system into one specific model, although the Polarized Pluralist model is probably the closest (McCargo 2012). However, the actual picture is too complex to describe the whole media system as one single model, although one cannot completely discard the applicability of this model for Bangladesh.

So far, the discussion renders the sharp differences between the two media systems reasonably explicit. Bangladesh's print journalism, with its rising circulation and critical editorial standpoint, is an influential element in the country's media world. In contrast, the print media in Australia, while unarguably influential, are accused of political bias. As a result, their circulation figures are declining. However, the picture is different in broadcasting, for whereas public service station ABC is a credible source of news and current affairs for many Australians, Bangladesh's state-run BTV is viewed as heavily biased towards the government of the day. For this reason, it has minimum credibility as a source of news. As some commentators have observed, BTV's strong editorial policy is directed towards serving "the state" (Chowdhury 2003), not the public. If one is to obtain a more detailed view of the two news media systems, it is necessary to examine their respective journalistic practices; for example, news content and the use of sources in news content. Paying specific attention to the products of journalism involves a shift in focus from generalising about the total media operations in a country (that is, using a single model to describe a country's diverse media organisations and behaviours) to viewing news media as separate and independent political organisations. Such a shift allows an appreciation of the differences between various news organisations within the media ecology of a single country. The salience of certain topics, or the use of sources in different news media, will facilitate an assessment of news media's diverse responses to crucial news and emerging issues. In this book, focus is upon examining news issues appertaining to climate change. Examination will provide an insight into how news media frame certain issues; for example, exercise a political position or exert independent influence on public debate (Schudson 2002) surrounding the macro-level environmental issue of climate change.

Conclusion

The above brief descriptions of the environmental issues and the political and media systems of Australia and Bangladesh reveal obvious differences between the two. In Australia, a century-long tradition of democratic practices has imbued its society with a certain character, different from that of Bangladesh. In terms of representative democracy, Bangladesh is at a very early stage, given that it has only been practising formal "electoral democracy" for a couple of decades. Apropos of geography, the divergence between the two countries is evident in two inherently different climatic conditions, i.e., the arid Australian landscape versus the low-lying, wet conditions in Bangladesh. The differences are also reflected in my brief review of the two countries' media systems, together with a distinctive similarity. There are marked differences in the ownership structures of their press systems, as well as in the modus operandi of their public service broadcasting systems. Differences are also evident in the ways in which members of the general public in the two countries perceive the credibility and value of these media operations. The similarity involves the difficulty of using a single conceptual tool (a specific model of media system) to describe the whole array of media operations within a country. While I do not suggest that Australia and Bangladesh have similar media environments, one faces considerable difficulty when using a single concept to explain various aspects of media practices (e.g., ownership patterns or perceptions of professional norms) in the two countries. This could potentially offer a nuanced understanding of the selection and representation of the various sources underpinning the coverage of climate change.

References

AC Nielson 2006, Bangladesh mass media and demographic survey, AC Nielson, Dhaka.Anam, M. 2007, 'Role of the media on democratic ownership of the development finance system', accessed October 11, 2009, available: www.oecd.org/data oecd/3/45/39364379.pdf.

Bacon, W. 2011, *A sceptical climate: Media coverage of climate change in Australia 2011*, Australian Centre for Independent Journalism, University of Technology Sydney.

Byrne, D. 2017, 'Australia's energy trilemma explained', accessed November 20, 2017, available: https://pursuit.unimelb.cdu.au/articles/australia-s-energy-trilemma-explained

Chadwick, P. 1996, 'Media ownership and rights of access', presented at the Free Speech Forum, hosted by the Communications Law Centre, Free Speech Committee of Victoria and Victorian Council for Civil Liberties at the State Film Centre, East Melbourne, December 8, accessed July 12, 2010, available: www.liberty victoria.org/speech-1996-chadwick.pdf.

Chadwick, P., Ferguson, S. & McCauslan, M. 1995, 'Shackled: The story of a regulatory slave', *Media International Australia*, vol. 77, August, pp. 65–72.

Chowdhury, A. 2003, *Media at the time of crisis: National and international issues*, Sraban Prakashoni, Dhaka, Bangladesh.

Commonwealth Scientific and Industrial Research Organisation (CSIRO) 2008, 'Water availability in the Murray-Darling Basin: A report from CSIRO to the Australian Government', accessed July 30, 2011, available: http://www.csiro.au/Organisation-Structure/ Flagships/Water-for-a-Healthy-Country-Flagship/Sustainable-Yields-Projects/WaterAva ilabilityInMurray-DarlingBasinMDBSY.aspx.

Commonwealth Scientific and Industrial Research Organisation (CSIRO) 2010, 'Groundwater threat to rivers worse than suspected', accessed July 15, 2011, available: http://www. csiro.au/en/Portals/Media/Groundwater-threat-to-rivers-worse-than-suspected.aspx.

Commonwealth Scientific and Industrial Research Organisation (CSIRO) 2015a, 'Climate change in Australia', accessed January 15, 2018, available: https://www. climatechangeinaustralia.gov.au/media/ccia/2.1.6/cms_page_media/168/CCIA_2015_ NRM_TechnicalReport_WEB.pdfhttp://www.csiro.au/en/Portals/Media/Groundwater- threat-to-rivers-worse-than-suspected.aspx.

Commonwealth Scientific and Industrial Research Organisation (CSIRO) 2015b, 'Climate change information for Australia', accessed January 25, 2018, available: https://www. csiro.au/en/Research/OandA/Areas/Oceans-and-climate/Climate-change-information.

Curran, J. 2002, *Power without responsibility: The press and broadcasting in Britain*, Routledge, London.

CunninghamS. & Flew, T. 2010, 'Policy', in S. Cunningham & G. Turner (eds) *The media and communications in Australia*, Allen & Unwin, Sydney, pp. 43–63.

Curran, J. & ParkM. J. 2000, *De-westernizing media studies*, Routledge, London.

Curran, J., Salovaara-Moring, I., Coen, S. & Iyengar, S. 2010, 'Crime, foreigners and hard news: A cross-national comparison of reporting and public perception', *Journalism*, vol. 11, no. 1, pp. 3–19.

Das, J. 2007, 'Sydney freelance journalists and the notion of professionalism', *Pacific Journalism Review*, vol. 13, no. 1, pp. 142–160.

Dasgupta, S., Khan, Z. H., Masud, S., Murshed, M., Ahmed, Z., Mukherjee, N. & Pandey, K. 2011, 'Climate proofing infrastructure in Bangladesh: The incremental cost of limiting future flood damage', *Journal of Environment & Development*, vol. 20, no. 2, pp. 167–190.

de Beer, A. & Merrill, C. J. 2004, *Global journalism: Topical issues and media systems*, Pearson, Boston, MA.

Department of Climate Change and Energy Efficiency 2011, 'Accurate answers to Professor Plimer's 101 climate change science questions', accessed August 30, 2012, available: http://www.climatechange.gov.au/climate-change/understanding-clima te-change/~/media/climate-change/prof-plimer-101-questions-response-pdf.pdf.

Diamond, J. 2005, *Collapse: How societies choose to fail or succeed*, Viking, New York.

Douglas, B., KearneyS. M. & Leatherman, S. P. (eds) 2001, *Sea level rise: History and consequences*, Academic Press, San Diego, CA.

Garnaut, R. 2011, *Garnaut review 2011: Australia in the Global Response to Climate Change*, Cambridge University Press, Port Melbourne, Vic.

Germanwatch 2012, 'Climate compatible development (CCD) in agriculture for food security in Bangladesh', accessed November 20, 2012, available: https://germanwa tch.org/en/download/8347.pdf.

Germanwatch 2018, 'Who suffers most from extreme weather events?', accessed March 20, 2018, available: https://germanwatch.org/en/download/20432.pdf.

Gocher, K. 2010, 'Volcano climate change', *ABC Rural*, Australian Broadcasting Corporation, accessed April 21, 2010, available: http://www.abc.net.au/rural/con tent/2010/s2878843.htm.

Griffin-Foley, B. 2003, *Party games: Australian politicians and the media from war to dismissal*, Text Inc., Melbourne, Vic.

Grove, R. H. 2007, 'The great El Niño of 1789–1793 and its global consequences: Reconstructing an extreme climatic event in world environmental history', *The Medieval History Journal*, vol. 10, pp. 43–46.

Gruen, F. & Grattan, M. 1998, 'Managing the pressure groups', in D. W. Lovell, I. McAllister, W. Maley & C. Kukathas (eds) *The Australian political system*, 2nd edn, Addison Wesley Longman Australia, Melbourne, Vic., pp. 394–402.

Gunaratne, A. S. 2007, 'Let many journalisms bloom: Cosmology, orientalism, and freedom', *China Media Research*, vol. 3, no. 4, pp. 60–73.

Gutierrez, M. 2016, 'Where does Bangladesh climate policy go, after Paris?', accessed December 22, 2017, available: https://cdkn.org/2016/08/feature-bangladesh-climate-p olicy-go-paris/?loclang=en_gb.

Hallin, D. C. 1996, 'Commercialism and professionalism in the American news media', in M. Gurevitch & J. Curran (eds) *Mass media and society*, Arnold, London, pp. 243–264.

Hallin, D. & Mancini, P. 2004, *Comparing media systems: Three models of media and politics*, Cambridge University Press, New York.

Hallin, D. & Mancini, P. 2012, *Comparing media systems beyond the Western world*, Cambridge University Press, New York.

Hallin, D. C. & Papathanassopoulos, S. 2002, 'Political clientelism and the media: Southern Europe and Latin America in comparative perspective', *Media, Culture & Society*, vol. 24, no. 2, pp. 175–195.

Hanitzsch, T. 2005, 'Journalists in Indonesia: Educated but timid watchdogs', *Journalism Studies*, vol. 6, no. 4, pp. 493–508. Harding-Smith, R. 2011, 'Media Ownership and Regulation in Australia', Centre for Policy Development, accessed October 28, 2012, available: http://cpd.org.au/wp-content/uploads/2011/11/Centre_for_Policy_Developm ent_Issue_Brief.pdf.

Henningham, J. 1996, 'Australian journalists' professional and ethical values', *Journalism and Mass Communication Quarterly*, vol. 73, no. 1, pp. 206–218.

Holmes, D. 2016, 'Why has climate change disappeared from the Australian election radar?', *The Conversation*, accessed October 10, 2018, available: https://theconversation. com/why-has-climate-change-disappeared-from-the-australian-election-radar-59809.

Hudson, M. 2018, 'The pro-coal Monash Forum may do little but blacken the names of a revered Australian', *The Conversation*, accessed April 18, 2018, available: https://the conversation.com/the-pro-coal-monash-forum-may-do-little-but-blacken-the-name-of- a-revered-australian-94329.

Huq, S. 2018, 'Updating Bangladesh's climate change strategy and action plan', accessed April 13, 2018, available: https://www.thedailystar.net/opinion/politics-climate-cha nge/updating-bangladeshs-climate-change-strategy-and-action-plan-1560916.

Independent Inquiry into the Media and Media Regulation 2012, *Report of the independent inquiry into the media and media regulation*, Department of Broadband, Communications and the Digital Economy, Canberra, ACT.

Inman, M. 2009, 'Where warming hits hard', *Nature Reports Climate Change*, 15 January, doi:10.1038/climate.2009.3, accessed September 18, 2012, available: http:// www.nature.com/climate/2009/0902/full/climate.2009.3.html.

Intergovernmental Panel on Climate Change (IPCC) 2007, 'General guideline on the use of scenario data for climate impact and adaptation assessment', Task Group on Data and Scenario Support for Impact and Climate Assessment (TGICA), IPCC, Geneva, Switzerland.

Irani, B. 2017, 'Life and obscurity of BTV', accessed April 5, 2018, available: https:// www.dhakatribune.com/opinion/special/2017/11/20/life-obscurity-btv/

Iyengar, S. 1991, *Is anyone responsible? How television frames political issues*, University of Chicago Press, Chicago, IL.

Jaensch, D., Brent, P., Bowden, B. 2005, 'Australian political parties in the spotlight', Australian National University, Canberra, accessed November 10, 2015, available: http://democratic.audit.anu.edu.au/papers/focussed_audits/200501_jaensch_parties.pdf

Jones, P. & Pusey, M. 2010, 'Political communication and "media system": The Australian canary', *Media, Culture & Society*, vol. 32, no. 3, pp. 451–471.

Josephi, B. 2005, 'Journalism in the global age: Between normative and empirical', *Gazette: The International Journal for Communication Studies*, vol. 67, no. 6, pp. 575–590.

Josephi, B. & Richards, I. 2012, 'The Australian journalist in the 21st century', in D. Weaver & L. Willnat (eds) *The global journalists in the 21st century*, Routledge, New York, pp. 115–125.

Karim, F. M. & Mimura, N. 2008, 'Impacts of climate change and sea-level rise on cyclonic storm surge floods', *Global Environmental Change*, vol. 18, pp. 490–500.

Khan, M. M. 2003, 'State of governance in Bangladesh', *The Round Table, The Commonwealth Journal of International Affairs*, vol. 92, no. 1, pp. 391–405.

Kochanek, A. S. 2000, 'Governance, patronage politics, and democratic transition in Bangladesh', *Asian Survey*, vol. 40, no. 3, pp. 530–550.

Kuhn, R. 1998, 'Who rules Australia?', in L. W. David, I. MacAllister, W. Maley & C. Kukathas (eds) *The Australian political system*, Addison Wesley Longman, South Melbourne, Vic., pp. 370–371.

James, R. M. 2010, 'The crisis in governance in two party systems', *Crikey*, accessed December 12, 2012, available: www.crikey.com.au

Jennings, K. 2012, 'Water under the bridge: A guide to the Murray-Darling Basin', in R. Koval (ed.), *The best Australian essays*, Black Inc., Melbourne, Vic., pp. 39–47.

Lawe-Davies, C. & Le Brocque, M. R. 2006, 'What's news in Australia?', in P. J. Shoemaker & A. A. Cohen (eds) *News around the world*, Routledge, New York, pp. 93–118.

Maddison, S. & Dennis, R. 2009, *An introduction to Australian public policy, theory and practice*, Cambridge University Press, Port Melbourne, Vic.

Marsh, I. 1995, *Beyond the two-party systems: Political representation, economic competitiveness and Australian politics*, Cambridge University Press, Melbourne, Vic.

Marsh, I. 2010, 'It's the system, stupid: Why Australia's two-party system has passed its use by date', *Crikey*, accessed November 24, 2012, available: www.crikey.com.au

McCargo, D. 2012, 'Partisan polyvalence: Characterizing the political role of Asian media', in D. Hallin & P. Mancini (eds) *Comparing media systems beyond the Western world*, Cambridge University Press, Cambridge, pp. 201–223.

McEachern, D. 1991, *Business mates: The power and politics of the Hawke era*, Prentice-Hall, Sydney, NSW.

McKnight, D. 2005, 'Murdoch and the culture war', in R. Manne (ed.) *Do not disturb: Is the media failing Australia?*Black Inc., Melbourne, Vic., pp. 53–74.

McQuail, D. 1994, *Mass communication theory: An introduction*, Sage, London.

Media Watch 2010, 'Spinning the science', Australian Broadcasting Corporation, accessed November 23, 2012, available: http://www.abc.net.au/mediawatch/transcripts/s2813774.htm.

Media Watch 2012, 'What's in a name', Australian Broadcasting Corporation, accessed December 20, 2012, available: http://www.abc.net.au/mediawatch/transcripts/s3458728.htm.

Merrill, J. C. (ed.) 1983, *Global journalism: A survey of the world's mass media*, Longman, New York.

Merrill, J. C. 2004, 'Global press philosophies', in A. de Beer & J. C. Merrill (eds) *Global journalism: Topical issues and media systems*, Pearson, Boston, MA, pp. 3–18.

Nash, C., Chubb, P. & Birnbauer, B. 2009, 'Fighting over fires: Climate change and the Victorian bushfires of 2009', paper presented at the Global Dialogue Conference, Aarhus, Denmark, November 3–6.

Nicholls, R., Hoozemans, F. M. J. & Marchand, M. 1999, 'Increasing flood risk and wetland losses due to global sea-level rise: Regional and global analyses', *Global Environmental Change*, vol. 9, no. 1, pp. 69–87.

Organisation of European Cooperation and Development (OECD) 2003, 'Development and climate change in Bangladesh: Focus on coastal flooding and the Sundarbans', accessed June 20, 2009, available: www.oecd.org/env/cc/21055658.pdf.

Parkinson, G. 2018, 'Coalition back-bench's crazy last gasp attempt to save coal', accessed April 19, 2018, available: https://reneweconomy.com.au/coalition-back-benchs-crazy-last-gasp-attempt-to-save-coal-76069/.

Parnini, N. S. 2006, 'Civil society and good governance in Bangladesh', *Asian Journal of Political Science*, vol. 14, no. 2, pp. 189–211.

Plimer, I. 2009, *Heaven and earth: Global warming: The missing science*, Conner Court Publication, Ballan, Vic.

Potter, E. & McKenzie, S. 2007, 'Introduction', in E. Potter, A. Mackinnon, S. McKenzie & J. McKay (eds) *Fresh water: New perspectives on water in Australia*, Melbourne University Press, Carlton, Vic., pp. 1–7.

Rahman, A. 2007, *Market-orientation in the news production of private TV channels in Bangladesh: An analysis in political economy approach upon NTV*, Masters Research Thesis, Department of Mass Communication, University of Rajshahi, Bangladesh.

Rahman, A. 2009, 'A political economy of the emerging television news industry in Bangladesh', *Revista de Economia Politica de las Tecnologias de las Informacion y Comunicadion*, vol. 11, no. 2, accessed April 19, 2011, available: www.eptic.com.br.

Rahman, M. 2005, 'When media takes a stand for social change: The experience of Prothom Alo in Bangladesh', accessed November 22, 2010, available: http://www.rmaf.org.ph/Awardees/Lecture/LectureRahmanMat.htm.

Ramaprasad, J. & Rahman, S. 2006, 'Tradition with a twist: A survey of Bangladeshi journalists', *The International Communication Gazette*, vol. 68, no. 2, pp. 148–165.

Rashid, H. & Paul, B. 2014, *Climate change in Bangladesh: Confronting impending disasters*, Lexington Books, Plymouth.

Roy, S. 2006, *Media map and value chain of media industry in Bangladesh: Special focus on SME's*, Media Professionals Group, Dhaka, Bangladesh.

Saul, B., Sherwood, S., McAdam, J., Stephens, T. & Slezak, J. 2012, *Climate change & Australia, warming to the global challenge*, Federation Press, Leichardt, NSW.

Schudson, M. 2002, 'The news media as political institutions', *Annual Review of Political Science*, vol. 5, pp. 249–269.

Siebert, F. S., Peterson, T., & Schramm, W. 1956, *Four theories of the press*, University of Illinois Press, Urbana, IL.

Tiffen, R. & Gittins, R. 2009, *How Australia compares*, Cambridge University Press, Cambridge.

Tiffen, R. 2010, 'The press', in S. Cunningham & G. Turner (eds) *Media and communication in Australia*, Allen & Unwin, Crows Nest, NSW.

Tuchman, G. 1976, 'Telling stories', *Journal of Communication*, vol. 26, no. 4, pp. 93–97.

Wahlquist, A. 2008, *Thirsty country: Options for Australia*, Allen & Unwin, Crows Nest, NSW.

Wasserman, H. & de Beer, A. S. 2009, 'Towards de-westernizing journalism studies', in K. Wahl-Jorgensen & T. Hanitzsch (eds) *The handbook of Journalism Studies*, Routledge, New York, pp. 428–438.

Weaver, D. 2005, 'Who are journalists?' in H. de Burgh (ed.), *Making journalists: Diverse Models, global issues*, Routledge, London, pp. 44–57.

Whittington, J. & Liston, P. 2003, 'Australia's rivers', in *Year Book 2003, A report from Australian Bureau of Statistics (ABS)*, accessed February 20, 2009, available: http://ewater.com.au/archive/crcfe/freshwater/publications.nsf/827558d21061a2f2ca256f1500 11f4da/36a59028a989408dca257022002b2927/$FILE/ch14spa.pdf.

Wooldridge, S. A. 2009, 'Water quality and coral bleaching thresholds: Formalising the linkage for the inshore reefs of the Great Barrier Reef, Australia', *Marine Pollution Bulletin*, vol. 58, no. 5, pp. 745–751.

World Bank 2014, 'Turn down the heat: confronting the new climate normal', accessed January 22, 2018, available: http://documents.worldbank.org/curated/en/31730146824209 8870/Main-report.

World Bank 2018, 'DataBank', accessed April 2, 2018, available: http://databank.worldba nk.org/data/reports.aspx?source=2&series=EN.ATM.CO2E.PC&country=accessed.

Wright, D. K. 1976, 'Professionalism levels of British Columbia's broadcast journalists: A communicator analysis', *Gazette*, vol. 2, no. 1, pp. 38–48.

Xiaoge, X. 2009, 'Development journalism', in K. Wahl-Jorgensen & T. Hanitzsch (eds) *The handbook of Journalism Studies*, Routledge, New York, pp. 357–370.

Zafarullah, H. & Siddiquee, A. N. 2001, 'Dissecting public sector corruption in Bangladesh: Issues and problem of control', *Public Organization Review: A Global Journal*, vol. 1, pp. 465–486.

3 News sources, journalism and the study

This chapter provides an overview of the studies of sources in journalism, particularly the use of sources by journalists in the news text. Focus is also on the notions of discourse and framing, and how they facilitate an analysis of struggles for power through the presence or absence of sources in news content. It is plausible to argue for the claim that "sources make the news," although the sources are not able to shape the news in exactly the way they would like (Tiffen et al. 2013, p. 2; Ericson et al. 1989). In fact, journalists play a significant role in the shaping and framing of news by including or excluding sources and content. However, sources are the crux of the idea of news as knowledge, for news workers must use them in every instance to establish the legitimacy of their content (Park 1940/2006). The balance of defining power between the sources and journalists lies in the subtle fact that while the journalists must use the sources, the news workers are able to decide in which manner they present the materials provided by the sources. These selection and presentation abilities enable journalists to exercise their performative power or influence (Broersma 2010). As a result, they get engaged in the contestation over the definitions of public issues and boundaries of public debates. In this contestation, the media professionals decide which sources' information become dominant or visible and what remain invisible or relatively less visible (Broersma et al. 2013). This chapter emphasises the visibility of sources as a crucial indicator of the pattern of source dominance in news pertinent to climate change in Australia and Bangladesh. This examination of the pattern of source dominance will highlight how representations of particular topics in the news are formulated, and how the attribution of responsibility, which is an important news frame, is influenced (Iyengar 1991). The chapter also discusses some methodological details, including the reasons for making various selections.

Voices in news content

Whose voice is present or absent in news content remains a crucial point in the analysis of journalism. The presence or absence is examined in this chapter by analysing news sources. A source, that is, a provider of information relevant to news content, can be an individual, a group, an organisation or other materials. Without sources, "modern news is unimaginable" (Carlson & Franklin 2011, p. 1). Journalists use sources for various practical reasons including incorporating witness observations of news events (direct quotes), providing validity to news articles through verification of accounts with the statements from "authorised knowers" (Tuchman 1978) and illustrating competing arguments (Dimitrova & Strömbäck 2009, p. 76). Critics argue that in the presentation of competing arguments:

> News is a representation of authority. In the contemporary knowledge society news represents who are the authorized knowers and what are their authoritative versions of reality. As such, it is every person's daily barometer of 'the knowledge-structure of society.' It offers a perpetual articulation of how society is socially stratified in terms of possession and use of knowledge.
>
> (Ericson et al. 1989, p. 3)

News is, in fact, a symbol of authority because it prioritises socially sanctioned knowledge about different issues as well as the views of "authorized knowers" holding significant positions in various social institutions. However, if journalism is considered to be the principal driving force of the public sphere (Habermas 2006), it is essential to look not only at authorised knowers, but also at the whole spectrum of sources of news irrespective of the sources' positions in the hierarchy of social power. Such comprehensive scrutiny would enable determination of who possesses the power to define news and who do not (Cottle 2000; Schlesinger 1990). The imbalance of power among the providers of information to journalists has always been the crux of the study of news sources (Bell 1994; Franklin et al. 2010). In the sociology of news, issues pertaining to this imbalance have been examined through the extremely complex web of interactions between the sources and journalists, which are manifested in what they interact about and how their interactions are structured.

Contestation: sources vs. journalists

When attempting to decipher this "extremely complex" web of interactions, scholars have focused on who is powerful in the negotiation of producing news. Indeed, the issue of power has dominated the debate surrounding source–journalist relationships for more than three decades (Tunstall 1971; Tuchman 1978; Gans 1979). Within this debate, many critics have argued that the interactions between the two sides are built on a reciprocal relationship in which journalists seek information, and sources, in turn, seek access to news media. There is mounting evidence to suggest that this relationship is symbiotic albeit in most cases, the sources enjoy an advantageous position over the journalists and lead the interactions (Sigal 1973; Nord & Strömbäck

2003). Critics adopting a sociological and political communication approach (Manning 2001; Bennett 2003) suggest that the newsroom culture, especially the deadline-driven news production process, renders the journalists particularly source-dependent (Strömbäck & Nord 2006, pp. 148–149). This process, which of necessity requires information-rich sources at short notice, puts the sources in an advantageous position.

However, those who disagree with this view argue that what occurs in the interaction is not a unilateral domination by sources but an ongoing "negotiation of newsworthiness" between journalists and sources (Ericsson et al. 1989; Cook 1998). During this negotiation, both the sources and the journalists attempt to control key information resources, and the outcome is not pre-determined in favour of the sources. "Acting as gate-keepers, journalists are in control of visibility, the extent to which sources should get the attention that they are seeking, and the tone of the news stories" (Strömbäck & Nord 2006, p. 148). On the other hand, the sources are in control of information, i.e., the power to decide what type of information they would divulge. Also, powerful sources have the capacity to legitimise a particular version of a news story. Thus, both parties come to this relationship possessing a particular power (control over visibility or information) and with a certain vulnerability (seeking information or visibility). This relationship engenders negotiations that are carried out by different actors from both sides to determine what a news story should be about and when and how a story should be made public. The "negotiation of newsworthiness" occurs at different stages; for example, in the process of news making, that is, negotiations concerning "where and when interactions [between the sources and journalists] will occur," and in the concern regarding the content of a news story. In other words, "what the story will be about and how it should be framed" (Cook 1998, p. 250). However, to date, critics have paid little attention to how journalists position themselves strategically against any powerful institutional or ordinary sources in any cross-national context. Most of the scrutiny in this area has been conducted in the context of Western media systems, particularly in the US (Josephi 2005). This has rendered the extant body of literature particularly inadequate for understanding the contestation of power between sources and journalists in the non-Western contexts (Waisbord 2010). In effect, the literature warrants an effective comparison of source–journalist interactions between the Western and non-Western contexts, leaving the similarities and differences between them inadequately understood. The contestation for dominance among the various sources and the degree of leverage they are able to garner in the production of news is useful for understanding the representation of sources in which the news media both support and, at times, challenge the sources. In this chapter, it is assumed that both support and challenge stem from the contestation between the sources and journalists. It would thus be more productive to examine different sources' roles in news content and the debate surrounding said roles. In the following section, the focus is upon a number of significant actors who are involved in the construction of news, e.g., politicians, bureaucrats, experts and activists.

Politicians and officials as dominant sources

Habermas (2006) views politicians and journalists as two important actors in the public sphere; without these actors, the public sphere is deemed to be inoperable. Although the above two represent the core, there are also at least five other types of actors who operate in the public sphere. These include lobbyists representing the interests of certain groups, experts credited with skills and knowledge, moral entrepreneurs calling public attention to a supposedly neglected matter, advocates representing marginalised groups, and intellectuals who have gained personal recognition for their contributions to particular fields. Schlesinger (2009), who disagrees with this view, maintains that taking politicians and journalists as the main actors signals a media-centric approach to the examination of source–media relationship. Rather, he suggests focusing on macro-level interfaces between the government, media and other civil society organisations. This will enable the identification of the communicative strategies of various actors (bureaucrats, experts, activists, etc.) in the contested world of news making. Despite the significance of the different voices in the news media, the platform from which information is disseminated remains heavily dependent upon political and official sources for raw material of news. This dependence raises questions about the quality of information in any democratic communication system (Bennett 2009). This dependence can be explained through the journalistic process of verification and, in particular, how verification works under deadline pressure. As Manning (2001) asserts:

> The pressure of news deadlines and the importance of obtaining information rich in news values encourage dependence upon official sources, whether they be government departments, sources associated with parliaments and formal policy-making process, the police or the other state control agencies.
>
> (p. 55)

The official sources to which Manning alludes are credible because the information they provide can be published without further research into its veracity (Ericson et al. 1989). Although this reliance on official sources risks reproducing the views of the powerful actors only, the journalistic conventions of news collection continue to rely on the "hierarchies of credibility" which emanate from the authority of source institutions (Becker 1967; Ericson et al. 1987). The relentless production of news round the clock is possible because the official authoritative institutions generate a huge amount of information into which news media are able to tap as raw materials (Tiffen 1978). Thus, the relationship between politicians and journalists is unequal because the latter are significantly reliant on the former and other official sources for their raw materials. The adversarial role of journalists requires reporters to ensure the credibility of their content by testing and corroborating the assertions of dominant sources (e.g., politicians) along with perspectives from other sources (e.g., the experts). As van Dijk argues, the "media tend to use 'experts' whose reputations and qualifications add weight to the argument being made, influence the way events are interpreted, and set the agenda for future debate" (quoted in Rowe et al. 2004, p. 161; also in Boyce 2006).

Experts as the "efficient machinery of record"

Few studies have shed light on how journalists use news sources to inform the citizenry and to what extent they do this responsibly and appropriately. Critics (Bell 1973; Giddens 1990; Albaek et al. 2003; Ericson et al. 1989) have greatly emphasised the significance of experts in modern knowledge societies due to increased use of scientific and technological knowledge. Similarly, Giddens (1990) in his book titled *The Consequences of Modernity* writes that modern societies are becoming increasingly reliant on "highly specialised expert systems" not only for solving their problems, but also for appreciating the complexities of the modern world. This is one of the reasons why journalists need experts as sources to assist them with information and interpretation of the day's issues. Drawing upon Walter Lippmann, Schudson (2006) examines the position of experts in journalistic practices and claims that citizens do not receive a picture of the world from the news content in a straightforward way; rather, they perceive it through the stereotypes provided by the press. If journalists depend upon trustworthy experts as an "efficient machinery of record," news content becomes more reliable. The importance of experts' representations of different issues in the news is evident here. However, it is essential to define the term "expert" clearly. For Lipmann (1922), experts are those who try to "put aside their own interests and wishes when they examine the world were the best hope to save democracy from itself" (Schudson 2006, p. 492). Schudson, on the other hand, places more emphasis on the skills and esoteric knowledge of the experts, which allow them to be recognised as such in the wider society as well as in their profession. But, the problem is how can experts maintain their independence in the face of potential pressure? Schudson (2006) observes that:

> Every governmental use of expertise is ultimately under the control of democratic authorities ... Actual problems about expertise in democracy are generally of two sorts. First, what are the best institutional mechanisms for keeping experts responsible to the people's representatives—while still enabling their expertise to bear on and improve decision-making? ... Second, how does democratic authority give experts enough autonomy so that the voice of the expert represents the experts' expertise rather than the views of the politicians or bureaucrats who pressure the experts into submission?
>
> (p. 497)

When experts profess their loyalty to the powerful rather than to their professional ethics and standards, this may lead to "group-think" pressure (Trembath 2010) within organisations that prevents the experts from freely and fairly representing their views. Drawing upon the Abu Ghraib prison torture incident in Iraq in 2003–2004, where the medical doctors were well aware of the torture inflicted on the prisoners by their captors, it may be asserted that experts often provide only the advice the politicians want to hear, and misrepresent the issue (Schudson 2006).

In terms of the implications of the use of expert sources in news, it is pertinent to discuss the relative importance of experts as against other sources. When investigating journalists' reasons for using experts as sources and the nature of the interactions between them in three Danish newspapers, Albaek and co-authors (2003; also Albaek 2011) identified a recent shift in the news production process which rendered "journalism by the journalist" more powerful than journalism by other news makers. As a result of this shift, journalists are now more prominent in the news content than individuals from dominant powerful groups, e.g., politicians, business executives and professionals (Albaek et al. 2003). In other words, compared to the above groups of individuals, journalists occupy considerable room in the news. Simultaneously, the practice of journalism is also shifting from descriptive (i.e., mainly authoritative sources describing what is going on) to more interpretative and investigative narratives in which the journalist employs diverse expert sources to ascertain not only *what* has happened, but also *why*. Albaek and co-authors also reveal that the use of sources in news content increased sevenfold between 1961 and 2001, similar to Patterson's (1991) findings in the US. Also, there has been the concomitant effect of the meteoric rise of sources, particularly expert sources; thus, journalism as a practice is becoming more independent from politicians and other news makers. This independence is reflected in the journalists' capacity to show scepticism to politicians and to replace "political logic" with "media logic." The media logic infers that "the requirements of the media take centre stage and shape the means by which political communication is played out by political actors, is covered by the media, and is understood by the people" (Albaek 2011, p. 337). Research has also revealed that selected experts or scientists seldom focus on or speak about their research results in the news media. Rather, they become increasingly involved in the political debates of the day. In the Danish news media, for example, scientists as expert sources are outnumbered by social scientists who have been chosen to speak, not for their expertise in particular fields, but for the general public's interest in them and in particular issues. This raises questions regarding how news media select their so-called "expert sources" to inform the readership. Irrespective of journalistic reasons for the selection of expert sources and the suitability of these sources in informing the citizenry, many critics (Weiler 1983; Giddens 1990; Schudson 2006) agree that there is room for experts in democracy who may have the capacity to "speak truth to power," that is, to clarify issues of public debate to both the politicians and the people and identify injustice in policy-related questions. This role of the experts can empower the general public. Viewed from this perspective, it is deemed necessary for this study to explore the position and representation of experts and scientists in the news coverage of climate change issues in the four selected newspapers.

Activists and the rise of media management

Concomitant with the rise of science and technology in modern knowledge societies, it is not only scientists and experts who are actively enhancing their presence in the public sphere; other organised and unorganised groups pushing their respective agendas are becoming more visible than ever before. Focusing on environmental issues, Hansen (2011) identifies the rise of different source groups as a "dialectical principle" in the public sphere, groups comprising different voices and counter voices espousing their views on multiple local and global environmental issues. Growing commercial pressure on journalism is also reflected in the shrinking news hole, and in increased media competition, which has increased the visibility of environmental issues as well as activists as sources receiving enhanced media attention. Sources do not confine themselves to attracting media attention alone; they also attempt to, and in many cases successfully establish the legitimacy of their claims. Critics (Solesbury 1976; Hansen 2011, p. 12) argue that the process of claiming legitimacy and invoking action operates according to the "dialectic principle in the public sphere" in which every claim prompts a counter claim (Castells 2004; Forde 2017). This principle often achieves extraordinary momentum due to the invocation of journalistic norm of balance in which the news media employ sources from both sides of a controversy or debate. Signitzer and Prexl (2007) argue that environmental activism has almost redoubled the activities of various business corporations to encounter, debate and undermine the claims of environmental pressure groups in the Australian news media. This finding is consistent with an earlier investigation by Sharon Beder (2002), who reveals how different corporate businesses in Australia engaged up-front activist groups to promote their interests in the public sphere. Similarly, Davis (2008) comments on the meteoric rise of public communication professionals across the world, which has changed the balance of power between sources and journalists. Public communication professionals act proactively and initiate "access by proxy" or "third party endorsement" of their messages through the news production process to manage the content to their desired directions. This development has blurred the identity of experts in the news, making it often difficult to differentiate between public communication professionals and actual experts in debates surrounding the public issues.

Several other studies undertaken in the US and Australia (Antilla 2005; Boykoff & Boykoff 2004, 2007; Gelbspan 1998, 2004, 2005; Leggett 2001; Lahsen 2005, in Antilla 2005, p. 340; McKnight 2010; Nash et al. 2009) demonstrate the implications of the selection of sources and of the particular positioning of them in the coverage of environmental issues including climate change. These investigations commonly find frequent usage of climate scep-tics, who are associated with the fossil fuel industry in the US and mining industry in Australia, as sources of news. And, on occasion, these sceptics become the dominant sources. While selection of these sources may fulfil the journalists' professional requirement of balance, it significantly undermines the premise of anthropogenic climate science by creating a false notion of mechanical "balance" in the media (Zehr 2000; Boykoff & Boykoff 2004). Scholars have argued that the social role of the news media is critical to understanding such positioning of sources. The above investigations reveal that anti-climate change sources are frequently portrayed in ways that allow them to present their "invalid, cynical and unsubstantiated" viewpoints to legitimise the perceived "phoney controversy" surrounding anthropogenic climate science (Antilla 2005). An important point to note here is journalists' power of selection. On a day-to-day basis, journalists possess the basic power to include certain sources in and exclude others from the production process. From the above discussion of the false balance in environmental news, it is fair to suggest that this decision-making power is absolutely crucial to the contestation of power in the symbolic field of news media. The selection and use of sources can lead to both symbiotic and adversarial relationships between the sources and journalists, depending upon the constantly shifting contexts of news making and structures of broad social power. The relation-ship between journalists and environmental activists as sources has been labelled by critics as a "ceaseless dance" (Hutchins & Lester 2015, p. 338) as it has been taking diverse turns in the reporting of climate change (Forde 2017; Waisbord 2018). The range of views about various sources offers a cri-tical insight into the disposition of sources in different national contexts. But, the above brief review also makes clear that cross-national comparison of sources has yet to receive adequate scholarly attention.

Comparison of sources

The above discussion has focused on the significance of various sources in the interactions between news media organisations and other institutions in society. Studies have highlighted the role of sources either as "testing grounds" for scrutinising powerful institutions in society or as "conveyor belts" for those institutions to convey their messages. However, these studies in the main have been limited to the perspectives of a single country. For this reason, the relevance of the above studies is somewhat limited for the current discussion as it seeks a meaningful comparison of source dominance and source–journalist interactions in different national contexts. As suggested in Chapter 1, very little attention has been paid to comparing sources in cross-national contexts. Given that the current study's focus is upon comparisons of sources, the following section explores studies that compare news sources across national contexts. Despite this lack of adequate attention, it is possible to identify a number of studies which have specifically dealt with the use of sources in news content in different countries. These studies have examined the issue of sources predominantly in relation to the effects of globalisation (Kim & Weaver 2003) and the dominance of official sources (Traquina 2004; Dimitrova & Strömbäck 2009). Kim and Weaver (2003) compared patterns of news sourcing in the coverage of the 1997 Asian Economic Crisis and the International Monetary Fund (IMF) bailout in five countries (the US, Korea, Indonesia, Thailand and Malaysia). They found that corporate sources and economic analysts were greater in number in the American media compared to media in other countries, whereas government sources were predominant in the Asian media. Similar to the Kim and Weaver study, analysis of sources of HIV/AIDS-related news in four countries (the US, Portugal, Spain and Brazil) revealed an overwhelming dominance of official sources, including government and World Health Organisation officials who were responsible for medical and scientific matters (Traquina 2004). The dominance of official sources was further evidenced in Dimitrova and Strömbäck's (2009) comparison of the use of sources in the US and Swedish newspaper coverage of three particular issues: war, elections and global controversy. The authors examined the 2003 Iraq War; the 2002 national elections in Sweden; the 2004 presidential election in the United States; and the global controversy surrounding the Mohammad cartoons in 2005/2006. They found official sources as more dominant in Sweden than in the US. A more recent study of the use of news sources in the 2009 European parliamentary election coverage in twelve countries (Austria, the Czech Republic, Denmark, Finland, Germany, Italy, Poland, Portugal, Romania, Spain, Sweden and the United Kingdom) also highlighted the importance of official sources, but particular focus was upon the influence of sources on the framing of politics in news reporting (Strömbäck et al. 2013). This issue of framing will be discussed further in later sections of this chapter.

Tiffen and colleagues (2013) have examined the "variations in news source patterns" across different news organisations in eleven countries. They intended to identify the diverse ways in which various political cultures had been using news sources and to understand the role of news media in creating an informed citizenry. The study examined print, television and online news content. As well, the authors conducted surveys of audience members in Australia, Canada, Colombia, Greece, India, Italy, Japan, Norway, South Korea, the United Kingdom and the United States to gauge the peoples' knowledge of public affairs. Their study found public service broadcasting in Italy was heavily reliant upon government official sources, and its commercial media were diverse in source use. In contrast, the public service news media in the UK, Australia and India used more diverse news sources than their respective commercial counterparts. Drawing on Curran & Park's (2000) de-Westernising media theory, Tiffen et al.'s study not only challenged the impact of globalisation and convergence in news practices, but argued that the use of sources was more heterogeneous than homogenous. The above studies provide a general overview of the prevailing scenario in terms of source use across the world, and they clearly demonstrate the dominance of official sources in the political and economic news. However, there is a paucity of studies comparing the use of news sources in the coverage of a scientific policy debate (e.g., environmental issues) in diverse national settings. The above studies only examine the types of news sources; they do not throw light on the processes and rationales for source selections in the "manufacturing" of news.

Verification of sources

Critics have indicated that all statements made by sources are questionable. According to Tuchman, it is possible to empirically examine the degree of journalistic scrutiny by measuring the extent to which journalists verify claims made by the sources (1978, p. 84). Verification is entrenched in the professional ideal of journalism, which is oriented towards establishing the "truth" or "fact."

> Truth-telling is a virtue in part because, as with so many moral matters, there's a strong element of reciprocity: even the most accomplished liars prefer not to be lied to or deceived. Thus, dishonesty is proscribed in most cultures, and areas where truth-telling standards are relaxed, are more or less clearly defined; story-telling art, entertainment, advertising and public relations, talk radio, and the blogosphere. Mainstream journalism, so called, is not one of them ... Indeed truth-telling per se is not enough. Good journalism must also be vigilant in order to expose, wherever possible, lies, evasions, deceptions, omissions and bullshit.
>
> (Scheuer 2008, p. 62)

Thus, it may be argued that the vigilant function of journalism through the exposure of wrongdoing in society is to represent independent reality (Lichtenberg 1995). However, as Hannah Arendt (1977, p. 44) contends, the "reality is different from, and more than, the totality of facts and events, which, anyhow, is unascertainable." Journalists can only inform society about certain ongoing issues, not the whole of reality. All knowledge claims are partial and thematise one or a few of all possible subjects; but, even a partial representation of reality requires journalists to have "a consistent method of testing information" and a "discipline of verification" to assess the reliability and veracity of information or interpretation (Kovach & Rosenstiel 2007, pp. 74, 77, & 82; Swain & Robertson 1995). Verification of facts is one of the central functions that journalists perform based on their professional commitment to truth. This verification is important, particularly when journalists are engaged "in competition with other powerful mediating institutions that produce information for persuasion or manipulation" (Kovach & Rosenstiel 2007, p. 81).

As discussed earlier, professional claims to accuracy of the representation of reality are heavily reliant on the extent of reporters' and editors' verification of different sources. Reporters are perceived here as the "providers of truth, or if not of truth, at least a reasonably reliable version of the facts" (Swain & Robertson 1995, p. 15). Trustworthy versions of the facts are crucial for journalism wherein reporters are geared to deadlines and required to identify facts promptly to establish the "facticity" (Tuchman 1978). Tuchman claims that in news reporting, verification of facts is both a "political" and "professional accomplishment" (1978, p. 83). In line with professional requirements, reporters contact sources; but, the selection of eligible sources for verification is a political decision. Selection is associated with the fundamental structural issues of journalism because it allows journalistic mastery and control that are the crux of good journalism, and helps to establish the legitimacy of "journalism's authority in the community" (Glasser & Marken 2005, p. 264). Notions of good journalism in particular underpin its watchdog role, manifested in investigative journalism that critically scrutinises the powerful, the aim being to expose any wrongdoing and enable journalism to win legitimacy as a central institution with far-reaching authority in a democratic society (Protess et al. 1991, quoted in Ekström et al. 2006, p. 292; de Burgh 2000; Ettema & Glasser 1998). Critics (de Burgh 2000; Ekström et al. 2010) have mainly examined the application of scrutiny or verification of the practice of investigative journalism because such reporting has a watchdog role which ensures political accountability. As well, it uses a considerable number of unnamed sources (Duffy & Williams 2011; Swain & Robertson 1995). Ensuring the scrutiny of published information strengthens the reliability of facts because, in the process of verification, journalists ask their sources a range of adversarial questions. Verification via a kind of cross-examination is essential for the accountability of the privileged and powerful (Clayman & Heritage 2002). Ideally, professional verification of sources (Ettema & Glasser 1998; Kovach & Rosenstiel 2007) should apply to all types of journalism. In controversial or highly technical matters, e.g., climate change, the question of source citation, selection of sources and verification of their claims is of vital professional importance.

Discourse and framing

Climate change involves risks that jeopardise the present and threaten the future. One of the aims of this discussion is to see how risks are constructed in climate change news from different media contexts. In doing so, the discussion will focus on the "power of media and communication in shaping cultural politics in the era of 'global mega-hazards' such as climate change" (Beck 1992, quoted in Carvalho & Burgess 2005, p. 1458). It will adopt discourse analysis as a suitable approach to examine newspaper text in diverse socio-political contexts because discourse analysis is able to capture disperse and often conflicting perspectives. Discourse analysis enables knowing and experiencing the world in a pervasive manner (Fairclough 1993; McGregor 2003), for discourses function as the means through which existing social relations are reproduced and challenged (Fairclough 1995). It is a strong tool for exercising power through the production and dissemination of knowledge. As well, discourses can be used for resistance and critique for its explicit and established connection to the language and exercise of power (Thomson 2005). The meaning of language manifested in the texts can be fixed or negotiated over time; this meaning contributes to the discursive narrative and influences ongoing considerations and actions in society. According to some critics, "the discursive constellations emergent through the text challenge or sustain unequal power relations" (Fairclough & Wodak 1997, Boykoff 2008). Similarly, van Dijk insists that the significance of discourse lies in its capacity to reveal "the (re) production and challenge of dominance" (van Dijk 1993, p. 249). Both these aspects—sustain or reproduction and challenge—have significant bearings on the news media representation of agents of social power (e.g., politicians, officials and experts). According to Fairclough (2003), discourses are:

> [W]ays of representing aspects of the world – the processes, relations and structures of the material world, the 'mental world' of thoughts, feelings, beliefs and so forth, and the social world. Particular aspects of the world may be represented differently, so we are generally in the position of having to consider the relationship between different discourses. Different discourses are different perspectives on the world, and they are associated with different relations people have to the world, which in turn depends on their positions in the world, their social and personal identities, and the social relationships in which they stand to other people.
>
> (p. 124)

This dynamic scope of discourses offers an open-ended opportunity to inspect and scrutinise contested meanings represented in media texts pertaining to climate change. An analysis of climate change discourses would facilitate the scrutiny of discursive struggles that contribute to the public knowledge concerning rising transboundary climate risks and pertinent political, economic and social issues. To facilitate this analysis, newspaper editorial texts may be selected as individual units of analysis because various discursive strategies stem from newspapers' views and news. For the purpose of the analysis in the next chapter, van Dijk's notion of persuasive strategies and Fairclough's intertextuality, or the idea that any text is linked in a chain of texts, reacting to, drawing in, and transforming other texts (Fairclough & Wodak 1997 p. 262), may be applied to examine climate change texts.

Besides the analysis of discourse, news texts have been subjected to scrutiny through the lens of framing. Critics (Carvalho 2007; Olausson 2009) have pointed out that as part of the social construction of the world, framing is the crux of discourses. Both the notions of framing and discourse complement each other. For this reason, this book focuses on the process of argumentation that builds up in editorial opinions through discourses and the framing of source excerpts. It is assumed that the notion of "framing" is essential for revealing the unfolding of climate change as a news issue. Following Erving Goffman, who maintains that we all actively classify, organize, and interpret our life experiences to make sense of them, Gamson considers that "Frame is a central organizing idea or storyline that provides meanings to events related to an issue. It is the core of a larger unit of public discourse or constituting elements of public discourse" (in Pan & Kosicki 1993, p. 56). These ideas relate to Gitlin (2003), who argues that in the case of news "[F]rames are principles of selection, emphasis and presentation composed of little tacit theories about what exists, what happens and what matters" (p. 6). This means the frame in news is not only ontological—what exists or what happens—but also ideological—what matters—in society; in other words, what is important to those who produce and consume news. This ideological issue of importance relates to a pecking order in news content when one thinks about framing. Thus, according to Fairclough (2003), "Framing also brings in questions about the ordering of voices in relation to each other in a text" (p. 62). Perhaps the ordering issue has prompted Carragee and Roefs (2004) to assert that the frames are "the imprint of power." Referring to renowned framing scholar Robert Entman, they maintain that frames demonstrate "the identity of actors or interests that competed to dominate the texts" (Entman 1993, p. 53). These competing interests may be analysed by scrutinising journalistic verification of sources excerpts as well as by raising some pertinent questions, such as: How does the news text prioritise these different interests? Whose interests are dominant and whose are negated or marginalised? What are the consequences or implications of these differences? The responses to these questions (described in Chapters 5 and 6) shed light on how discourses are utilised in the exercise of power between various voices.

Framing is also an outcome of discursive habit underpinned by journalistic professional norms, genres (or linguistic style) and ideological convention based on "taken for granted assumption." So, the construction of a frame or an argument is not an active promotion of certain powerful sources over others; rather, "it is a matter of deployment of culturally constructed codes that function ideologically, as if they were the creations of nature" (Olausson 2009, p. 423). By discussing framing in relation to sources, this book aims to demonstrate that the selection and positioning of sources in the four newspapers contribute to the portrayal of climate change debates in a manner that seemingly justifies certain positions pertaining to environmental debates, and may delegitimise others. The notion of framing also helps to establish how and why journalists offer salience to certain events and sources over others, and what processes they follow to give meaning to the various phenomena that emerge in the coverage of climate change (Tuchman 1978; Pan & Kosicki 1993; Entman 1993).

In this book, broadly two types of frames have been studied: the action frame and the conflict frame. Critics defined the conflict frame as a matter of contestation between two or more forces in a social context; the action frame emphasises an act intended to solve an urgent problem or address a critical issue. In the conflict frame, contesting groups or entities may remain distant; but, in the action frame, they would perhaps align their interests for the sake of resolving the problem or issue (Benford & Snow 2000). The examination of these frames helps to determine the extent to which they support or challenge various positions in the local and global climate change debates in the two countries. Schudson (2011), who views framing as an organisational process and product, argues that as an analytical tool, it "opens the discussion to examining unintentional (even unconscious) as well as intentional selective presentation. It (framing) diminishes the extent to which evidence of selection can be automatically read as evidence of deceit, dissembling, or prejudice of the individual journalist" (pp. 30–31).

This understanding of framing manifestation allows going beyond the reductionist perspective of news practice often manifested in the notions of balance, accuracy and objectivity (Reese 2007; Schudson 2011). As well, it illuminates the process of selection of certain sources over others and the extent to which journalists seek to verify all of their sources with similar rigour. This brief discussion of various studies of news sources and the significance of their representation highlights both the selection and presentation of sources, and builds a case for examining the journalistic capacity for mediating ongoing issues in the news and comparing this capacity in cross-national professional practices.

Methods

This section explicates the design of the study presented in the subsequent chapters, i.e., its aim, the selection and analysis of content (two countries, two newspapers from each country, and two periods of data collection). As discussed earlier, the interpretation of editorial opinions (in Chapter 4) is important for setting the context for analysing the presence or absence of various sources and their representations, discussed in Chapters 5 and 6. The sources will be analysed as objects of investigation by taking into consideration various contextual influences, such as Australian and Bangladeshi media systems.

One of the purposes of this study is to initiate a discussion about the comparative characteristics of journalistic practices in Western and non-Western countries. It assumes that the analysis of news content is a suitable way of examining different characteristics of news practices and of obtaining a meaningful comparison of journalistic practices between the two countries. The newspapers selected for this study are *The Australian* and *The Sydney Morning Herald* from Australia, and *The Daily Star* and the *Prothom Alo* from Bangladesh. The periods of data collection are 2009 and 2015. The reasons for these and other methodological selections are discussed below. The study provides a general context for climate change news by highlighting the differences and similarities in coverage throughout 2009 and 2015. News content was examined to identify the types of sources and cross-checking of attributed source statements for the issues raised in the climate change news content. These variables are explained later in this section and detailed in Appendix 2 (Tables 1–7).

Rationale for the choice of countries for comparison

This comparison of newspaper content between an economically advanced, industrialised, dispersedly populated "honorary Western country like Australia" (Curran & Park 2000, p. 3) and a poor, agriculture-based, densely populated Third World country like Bangladesh has proven quite challenging. As suggested in Chapter 2, the difference between these two countries is further emphasised by their somewhat contrasting geographical features: Australia is a dry, arid land, while Bangladesh is a low-lying wet country, aptly called in common parlance the "land of rivers." Comparative studies are usually conducted between generally similar cases that are contrastive on one or only a few variables. In the field of Journalism Studies, very little comparative research has been undertaken as it is in the early stages of development (Hanitzsch 2009). Among the existing comparative studies in this field, similarity of cases is common. In addition, West-centrism is generally well recognised in this field (Josephi 2005). But, West-centric studies limit the empirical breadth of the existing literature. Thus, a comparison of dissimilar cases may prove beneficial for a number of reasons. First, comparative inquiries in Journalism Studies need diversification so that the current West-centric empirical narrowness can be adequately questioned, challenged and expanded. Second, comparing similar cases has led to "ideal typification" (Curran et al. 2010) by coupling journalism with democracy (Carey 2007) and highlighting the nation-state as the foremost unit of analysis. Comparing dissimilar cases allows scrutiny of these taken-for-granted positions and assumptions. For the above reasons, the exploration of journalistic practices in two dissimilar regions is productive given that it allows what Downing terms "communication theorising to develop itself comparatively," to go beyond the universalisation of the experience of a few unrepresentative nations such as the UK and the US (Downing 1996, p. xi, in Curran & Park 2000, p. 3; Josephi 2005).

Justification for selecting print media

In this study, print media have been selected to examine journalistic practices in the two countries. The reason underpinning this selection is the continued importance of newspapers in the world of journalism despite the many changes brought about by newer technologies and subsequent developments. In many contemporary discussions, profound pessimism is expressed regarding the future of newspapers, with many identifying the printed newspaper as an "endangered species" in the rapidly evolving media universe. As a contributor to the *Economist* magazine commented, "the business of selling words to readers and readers to advertisers, which has sustained their role in society, is falling apart" (Economist 2006). However, some critics are less negative about the future of newspapers. Although they do not discount the declining circulation, they do not support warnings that newspapers will "vanish" anytime soon. Rather, in the face of stiff challenges, newspapers are adapting to the new media environment by changing their content, style and design to suit a highly competitive and fragmented market operating on the platform of mobile telephony and the Internet (Franklin 2008). These contrasting perspectives have resonated in various discussions about print media in Australia (Tiffen 2010; McKnight & O'Donnell 2011), particularly in relation to Fairfax or News Limited publications. While the "business model" of mass-circulated newspapers is under threat—particularly in the industrialised Western countries—due to a shift in advertising revenue or the so-called "rivers of gold" from print to online platforms (McKnight & O'Donnell 2011; see also Media Watch 2012), critics regularly dispute the notion that newspapers in Australia are turning into dinosaurs.

> These so-called dinosaurs also provide raw materials for radio talkback, TV news shows, other newspapers and the Twitterati ... Newspapers matters for other reasons. They and their journalists set the political agenda for electronic media and the Internet. Murdoch's Sydney tabloid *The Daily Telegraph* and *The Australian* are said to be the most influential news media in Australia ...
>
> (McKnight & O'Donnell 2011)

The above quote clearly identifies the significance of the print media industry in Australia as an influential information and opinion provider, which shares many features with similar industries in comparable societies and markets (Manne 2005). At the same time, the print media industry has some features which are quite distinct in particular countries or regions. The circulation trend is a case in point. Unlike Western industrialised countries, daily newspapers in many Asian countries have been experiencing a significant rise in circulation, particularly among the growing middle class and those who have limited access to broadband networks. Newspaper sales in Bangladesh, for example, increased by 30 per cent in the five years preceding 2006. This coincides with an Asia-wide surge in newspaper circulation (Campbell 2017), despite the poor literacy rate in many Asian countries (for example, in Bangladesh only 56 per cent of the population is literate).

Due to their healthy rate of growth, newspapers in Bangladesh continue to thrive and play a strong role in the country's journalism as well as in politics and society generally. One rarely hears newspapers in this country being alluded to as "dinosaurs" or "endangered species" or criticism that their business model is "falling apart." According to official statistics, there are 467 newspapers in Bangladesh including 292 dailies, but only a few dozen of these are substantial and regular publications with significant reader support. The continued rise in the circulation of newspapers means that Bangladesh's print media industry remains an important component of the country's journalism and deserves continued scholarly attention.

The selection of the four newspapers

The four major newspapers selected for this study have considerable bearing on their respective country's policy circles and are able to set the agenda of the day on crucial political issues. The similar positions of these newspapers in terms of prestige and influence in their respective countries render these publications "suitable for comparative analysis" (Benson & Hallin 2007). Over the years, these newspapers have meticulously scrutinised a range of environmental issues including problems surrounding climate change (Das et al. 2009). In Australia, *The Australian,* a widely circulated centre-right, national broadsheet was first published in 1964 from its headquarters in Canberra. *The Sydney Morning Herald,* the country's oldest continuously published newspaper, is a metropolitan publication. Many consider it to be "Australia's most important newspaper. It is one of the high-ranking oldest newspapers published from Sydney since 1831" (Isaacs & Kirkpatrick 2003). As of 2018, the weekday circulation of *The Sydney Morning Herald* was 78,798 (Samios 2018). *The Australian's* circulation figure for 2012 was 133,701 (Dyer 2012), but there was no updated figure for this publication because *The Australian's* parent company "News Corp [has] decided to withdraw from the newspaper circulation audit" (Samios 2018). However, the paper's website claims that the total monthly reach of the print newspaper is 1.9 million (Australian 2019).

The two Bangladeshi publications chosen for this study are broadsheet newspapers, based in the capital Dhaka, and privately owned by a leading media company. The English language broadsheet *The Daily Star,* which carries the slogan "Your Right to Know" in its masthead, was established by prominent editor Syed Mohammed Ali (S. M. Ali) in the early 1990s. It is now part of a large media company. The Bangla language daily the *Prothom Alo* was founded by its editor Motiur Rahman in the mid-1990s. The publication is now part of the same company which owns *The Daily Star.* In general, many newspapers in Bangladesh, including the two selected for analysis, follow an editorial middle ground in politics, both supporting and criticising the government of the day. Although both papers can be considered centre-left dailies in terms of political outlook, the *Prothom Alo* seems to adopt a more adversarial position regarding different social institutions (e.g., politics and businesses) than the other publication. Not only can it claim the largest circulation in Bangladesh, but its website is the most popular media site among local and expatriate online users. In 2018, its circulation figure numbered 501,800. At the same time, *The Daily Star* numbered a mere 44,814 (Department of Film and Publication 2018). The circulation figures for the English dailies in the two countries show a wide gap in circulation numbers. While *The Australian* enjoys an average circulation in excess 133,000 during the weekdays, *The Daily Star* barely exceeds 40,000. However, a few factors need to be considered when comparing these figures, for example population size, literacy rates and the official languages of these countries. Australia has a population of 25 million as against over 168 million in Bangladesh (United Nations Department of Economic and Social Affairs 2017). However, in both cases, population size should be measured against factors such as the overall high rate of literacy in Australia (where English is the most commonly used language) and the low literacy rate in Bangladesh (where Bangla is the official language). This will help to explain the wide gap in circulation. Nevertheless, the small circulation figure for *The Daily Star* does not fully reflect its importance and influence as a national newspaper in Bangladesh.

Why climate change?

Climate change as the issue of focus has been selected for its perceived significance for both Australia and Bangladesh due to their environmental vulnerabilities. In Bangladesh, the issue of climate change is mostly congruous; but, in Australia, it is heavily contested for its scientific and policy implications. In 2009, the media particularly focused on the issue of climate change. The UN Climate Change Summit convened in Copenhagen was expected to effectively address the biggest "diabolical policy problems" of contemporary times. The summit's aim was to bring together both the developed and developing nations to mitigate the levels of emissions and set reduction guidelines across the world. Climate change, which had emerged as an important global policy issue across various regions, was closely connected to complex questions of politics, economics and equity. For these reasons, it drew considerable attention from the media across the spectrum. In 2015, the global attention was again focused on the issue of climate change, albeit to a lesser degree, prior to the Paris summit, which achieved the required universal agreement on emissions reduction that was so elusive in Copenhagen.

Significance of the monitoring period

The selected periods were politically significant. Two new administrations were in power in both countries during 2009. In Australia, the Rudd Labor government was elected to power in 2007 after 11 years of John Howard's Conservative Coalition regime. In Bangladesh, Sheikh Hasina's Awami League government came in office in early 2009, replacing an interim bureaucratic caretaker administration. Initially, both new governments showed genuine interest in their respective countries' climate change issues. In the period leading up to the Paris Summit (COP21) in 2015, both high-income and low-income countries committed to reduce greenhouse gas emissions. The ambition to "net zero carbon" by 2050 means this commitment will intensify in the decades to come. It is critical in this context that we gain a fuller understanding of the dynamics of climate communication in low-emitting, low-income countries, such as Bangladesh, as much as the high-income, high-emitting countries, such as Australia. The two-year window was considered sufficiently wide to capture different important aspects of the news coverage of climate change, e.g., the presence or absence of different sources and their representations.

Comparative structure of the study

I conducted extensive data collection on climate change news in two phases. In the first phase, quantitative data were collected, while in the second phase a mix of quantitative and qualitative data was generated. Together, the data set has enabled this study to offer a comprehensive analysis of the pertinent issues. Altogether, 3,998 news articles on climate change that appeared in the four newspapers were collected and examined during the study periods (2009 and 2015). Eight variables were identified to generate the basic quantitative data required to obtain an adequate detail of the nature of the news coverage: principal source, political source, bureaucratic source, activist source, expert source, business source, citizen source and verification of sources' statements. Appropriate values were assigned to each variable to tease out patterns of emphasis or prevalence across the range. The findings in regard to these variables are described in Appendix 2.

Data collection

In terms of the actual collection of news content, both the database and manual searches were conducted depending on the availability of newspaper content. For the content of the two Australian newspapers, various search terms, such as "climate change," "Copenhagen" and "Paris" were used in the Factiva database during six months (July–December) in 2009 and 2015. Regarding the Bangladesh news content, the archives of the respective newspaper websites were manually searched using the same search terms as those of the Australian content for the same six-month periods.

Research questions

The above-mentioned variables, such as news topics and sources, were adopted to comprehend the patterns of journalistic practice in Australia and Bangladesh. In particular, the presence of various climate change topics was explored to reveal the degree of news media emphasis on certain issues over others. Furthermore, the observed patterns of selected sources were expected to illuminate the framing strategies adopted by journalists in relation to important environmental issues in the respective countries. To this end, a number of crucial questions were posed:

- How did the newspaper coverage represent climate change in Australia and Bangladesh?
- What are the similarities and differences between the newspapers' coverage of climate change issues?
- What are the types of sources cited in the articles selected for the purpose of this study?
- Who are the principal sources in the selected articles?
- Is there any difference between the presence of principal sources in the two countries?
- How are the principal sources positioned or framed in the news articles?
- How, and to what extent, do the journalists verify various assertions made by their principal sources?
- What are the differences and similarities in the ways in which articles are verified, and in the responsibility for environmental problems attributed to in the two countries?

These questions were designed to generate detailed discussions pertinent to the focus issues. The outcomes are presented in the subsequent chapters.

Conclusion

This chapter has provided a description of the two-step content analysis process (quantitative and qualitative) undertaken to examine newspaper coverage of climate change. The rationale for selecting different variables and study periods to generate data for understanding the journalistic practices in Australia and Bangladesh have also been delineated. Based on this analytical guideline, the following chapters present the findings of the analysis and consider their implications not only for comparative journalism, but for intranational and international coverage of climate issues during a controversial period in the politics of climate change debate.

References

Albaek, E. 2011, 'The interaction between experts and journalists in news journalism', *Journalism*, vol. 12, no. 3, pp. 335–348.

Albaek, E.Christiansen, M. & Togeby, L. 2003, 'Experts in the mass media: Researchers as sources in Danish daily newspapers, 1961–2001', *Journalism & Mass Communication Quarterly*, vol. 80, no. 4, pp. 937–949.

Antilla, L. 2005, 'Climate of skepticism: US newspaper coverage of the science of climate change', *Global Environmental Change*, vol. 15, no. 4, pp. 338–352.

Arendt, H. 1977, *Between the past and future: Eight exercises in political thought*, Viking, New York.

Australian, The 2019, 'About *The Australian*', accessed February 16, 2019, available: https://www.newscorpaustralia.com/brand/australian/.

Beck, U. 1992, *Risk society: Towards a new modernity*, Sage, London.

Becker, H. S. 1967, 'Whose side are we on?', *Social Problems*, vol. 14, no. 3, pp. 239–247.

Beder, S. 2002, *Global spin: The corporate assault on environmentalism*, Green Books, Totnes, UK.

Bell, A. 1994, 'Media (mis)communication on the sciences of climate change', *Public Understanding of Science*, vol. 3, pp. 259–275.

Bell, D. 1973, *The coming of post-industrial society*, Basic Books, New York.

Benford, R. D. & Snow, D. A. 2000, 'Framing processes and social movements: An overview and assessment', *Annual Review of Sociology*, vol. 26, pp. 611–639.

Bennett, W. L. 2003, 'Communicating Global Activism, Strengths and Vulnerabilities of Networked Politics', *Information, Communication & Society*, vol. 6, no. 2, pp. 143–168.

Bennett, W. L. 2009, *News: The politics of illusion*, 8th edn, Longman, New York.

Benson, R. & Hallin, D. 2007, 'How states, markets and globalization shape the news: The French and American national press, 1965–1997', *European Journal of Communication*, vol. 22 no. 1, pp. 27–48.

Boyce, T. 2006, 'Journalism and expertise', *Journalism Studies*, vol. 7, no. 6, pp. 889–906.

Boykoff, M. 2008, 'Media and scientific communication: A case of climate change', in D. G. E. Liverman, C. P. G. Pereira & B. Marker (eds) *Communicating environmental geoscience*, Geological Society Special Publications, London, 305, pp. 11–18.

Boykoff, M. T. & Boykoff, J. M. 2004, 'Balance and bias: Global warming and the US prestige press', *Global Environmental Change*, vol. 14, no. 2, pp. 125–136.

Boykoff, M. T. & Boykoff, J. M. 2007, 'Climate change and journalistic norms: A case study of US mass media coverage', *Geoforum*, vol. 38, no. 6, pp. 1190–1204.

Broersma, M. 2010, 'The unbearable limitations of journalism: On press critique and journalism's claim to truth', *International Communication Gazette*, vol. 72, no. 1, pp. 21–33.

Broersma, M., den Herder, B. & Schohaus, B. 2013, 'A question of power: The changing dynamics between journalists and sources', *Journalism Practice*, vol. 7, no. 4, pp. 388–395.

Campbell, C. 2017, 'World press trend 2017', World Association of Newspapers and News Publishers (WAN-IFRA), accessed December 12, 2018, available: http://anp.cl/wp-content/uploads/2017/10/WAN-IFRA_WPT_2017.pdf.

Carey, J. W. 2007, 'A short history of journalism for journalists: A proposal and an essay', *International Journal of Press/Politics*, vol. 12, no. 1, pp. 3–16.

Carlson, M. & Franklin, B. 2011, 'Introduction', in B. Franklin & M. Carlson (eds) *Journalists, sources and credibility, new perspectives*, Routledge, New York, pp. 1–5.

Carragee, K. M. & Roefs, M. 2004, 'The neglect of power in recent framing research', *Journal of Communication*, vol. 54, no. 2, pp. 214–233.

Carvalho, A. 2007, 'Ideological cultures and media discourses on scientific knowledge: Re-reading news and climate change', *Public Understanding of Science*, vol. 16, pp. 223–243.

Carvalho, A. & Burgess, J. 2005, 'Cultural circuits of climate change in U.K. broadsheet newspapers, 1985-2003', *Risk Analysis*, vol. 25, no. 6, pp. 1457–1469.

Castells, M. 2004, *The power of identity*, 2nd edn. Blackwell, Oxford.

Clayman, S. & Heritage, J. 2002, *The journalists and the public figures on the air*, Cambridge University Press, Cambridge.

Cook, T. E. 1998, *Governing with the news: The news media as a political institution*, University of Chicago Press, Chicago, IL.

Cottle, S. 2000, 'Rethinking news access', *Journalism Studies*, vol. 1, no. 3, pp. 427–448.

Curran, J. & ParkM. J. 2000, *De-Westernizing media studies*, Routledge, London.

Curran, J., Salovaara-Moring, I., Coen, S. & Iyengar, S. 2010, 'Crime, foreigners and hard news: A cross-national comparison of reporting and public perception', *Journalism*, vol. 11, no. 1, pp. 3–19.

Das, J., Bacon, W. & Zaman, A. 2009, 'Covering environmental issues and global warming in delta land: A study of three newspapers', *Pacific Journalism Review*, vol. 15, no. 2, pp. 10–32.

Davis, A. 2008, 'Public relations in the news', in B. Franklin (ed.) *Pulling newspapers apart: Analysing print journalism*, Routledge, London, pp. 272–281.

de Burgh, H. (ed.) 2000*Investigative journalism: Context and practices*, Routledge, London.

Department of Film and Publication 2018, 'Statistics on the country's newspapers and periodicals included in the media list', Audit Branch, Department of Film and Publication, Dhaka, Bangladesh.

Dimitrova, D. & Strömbäck, J. 2009, 'Look who's talking: Use of sources in newspaper coverage in Sweden and the United States', *Journalism Practice*, vol. 3, no. 1, pp. 75–91.

Downing, J. 1996, *Internationalising media theory*, Sage, London.

Duffy, J. M. & Williams, E.A. 2011, 'Use of unnamed sources drops from peak in 1960s and 1970s', *Newspaper Research Journal*, vol. 32, no. 4, pp. 6–21.

Dyer, G. 2012, 'Newspaper circulation carnage—Biggest March fall on record', *Crikey*, accessed October 11, 2012, available: http://www.crikey.com.au/2012/05/11/newspaper-circulation-carnage-biggest-march-fall-on-record/?wpmp_switcher=mobile.

Economist, The 2006, 'Who killed the newspaper?', August 24, accessed June 3, 2011, available: http://www.economist.com/node/7830218.

Ekström, M., Johansson, B. & Larsson, L. 2006, 'Journalism and local politics: A study of scrutiny and accountability in Swedish journalism', *Journalism Studies*, vol. 7, no. 2, pp. 292–311.

Ekström, M., Johansson, B. & Larsson, L. 2010, 'Journalism and local politics', in S. Allan (ed.) *The Routledge companion to news and journalism*, Routledge, London, pp. 256–266.

Entman, R. 1993, 'Framing: Toward clarification of a fractured paradigm', *Journal of Communication*, vol. 43, no. 4, pp. 51–58.

Ericson, R. V., Baranek, P. M. & Chan, J. B. L. 1987, *Visualizing deviance: A study of news organization*, University of Toronto Press, Toronto.

Ericson, R. V., Baranek, P. M. & Chan, J. B. L. 1989, *Negotiating control: A study of news sources*, Open University Press, Milton Keynes.

Ettema, J. S. & Glasser, T. L. 1998, *Custodians of conscience: Investigative journalism and public virtue*, Columbia University Press, New York.

Fairclough, N. 1993, 'Critical discourse analysis and the marketization of public discourse: The universities', *Discourse and Society*, vol. 4, no. 2, pp. 133–168.

Fairclough, N. 1995, *Media discourse*, Edward Arnold, London.

Fairclough, N. & Wodak, R. 1997, 'Critical discourse analysis', in T. van Dijk (ed) *Discourse as social interaction*, Sage, London, pp. 258–284.

Fairclough, N. 2003, *Analysing discourse: Textual analysis for social research*, Routledge, London.

Forde, S. 2017, 'Environmental protest, politics and media interactions', in R. Hackett, S. Forde, K. Foxwell-Norton & S. Gunster (eds) *Journalism and climate crisis*, Routledge, London, pp. 77–93.

Franklin, B. 2008, 'Newspapers: Trends and developments', in B. Franklin (ed.) *Pulling newspapers apart: Analysing print journalism*, Routledge, New York, pp. 1–36.

Franklin, B., Lewis, J. & Williams, A. 2010, 'Journalism, news sources and public relations', in S. Allan (ed.) *The Routledge companion to news and journalism*, Routledge, London, pp. 202–212.

Gans, H. J. 1979, *Deciding what's news*, Pantheon Books, New York.

Gelbspan, R. 1998, *The heat is on: The climate crisis, the cover-up, the prescription*, Perseus Book, Cambridge, MA.

Gelbspan, R. 2004, *Boiling point*, Basic Books, New York.

Gelbspan, R. 2005, 'Snowed', *Mother Jones*, June, pp. 42–43.

Giddens, A. 1990, *The consequences of modernity*, Stanford University Press, Stanford, CA.

Gitlin, T. 2003, *The whole world is watching: Mass media in the making and unmaking of the new left*, University of California Press, Berkeley, CA.

Glasser, L. T. & Marken, L. 2005, 'Can we make journalists better?', in H. de Burgh (ed.) *Making journalists: Diverse models, global issues*, Routledge, London, pp. 264–276.

Habermas, J. 2006, 'Political communication in media society: Does democracy still enjoy an epistemic dimension? The impact of normative theory on empirical research', *Communication Theory*, vol. 16, pp. 411–426.

Hanitzsch, T. 2009, 'Comparative journalism studies', in K. Wahl-Jorgensen & T. Hanitzsch (eds) *The handbook of journalism studies*, Routledge, London, pp. 413–427.

Hansen, A. 2011, 'Communication, media and environment: Towards reconnecting research on the production, content and social implications of environmental communication', *International Communication Gazette*, vol. 73, no. 1–2, pp. 7–25.

Hutchins, B. & Lester, L. 2015, 'Therorizing the enactment of mediatized environmental conflict', *International Communication Gazette*, vol. 77, no. 4, pp. 337-358.

Isaacs, V. & Kirkpatrick, R. 2003, *Two hundred years of Sydney newspapers: A short history*, Rural Press, North Richmond, NSW.

Iyengar, S. 1991, *Is anyone responsible? How television frames political issues*, University of Chicago Press, Chicago, IL.

Josephi, B. 2005, 'Journalism in the global age: Between normative and empirical', *Gazette: The International Journal for Communication Studies*, vol. 67, no. 6, pp. 575–590.

Kim, T. S. & Weaver, H. D. 2003, 'Reporting on globalisation: A comparative analysis of sourcing patterns in five countries' newspapers', *Gazette: The International Journal for Communication Studies*, vol. 65, no. 2, pp. 121–144.

Kovach, B. & Rosenstiel, T. 2007, *The elements of journalism: What newspeople should know and public should expect*, Three Rivers Press, New York.

Lahsen, M. 2005, 'Technocracy, democracy, and US climate politics: The need for demarcations', *Science, Technology, & Human Values*, vol. 30, no. 1, pp. 137–169.

Leggett, J. 2001, *The carbon war: Global warming and the end of the oil era*, Routledge, London.

Lichtenberg, J. 1995, 'In defence of objectivity revisited', in M. Gurevitch & J. Curran (eds) *Mass media and society*, Arnold, London, pp. 225–242.

Lipmann, W. 1922, *Public opinion*, Macmillan, New York.

Manne, R. (ed.) 2005, *Do not disturb: Is the media failing Australia?*, Black Inc., Melbourne, Vic.

Manning, P. 2001, *News and news sources: A critical introduction*, Sage, London.

McGregor, S. 2003, 'Critical discourse analysis – a primer', *Kappa Omicron Nu Forum*, vol. 15, no. 1, accessed November 11, 2018, available: http://www.kon.org/archives/forum/15–1/mcgregorcda.html.

McKnight, D. 2010, 'A change in the climate? The journalism of opinion at news corporation', *Journalism*, vol. 11, no. 6, pp. 693–706.

McKnight, D. & O'Donnell, P. 2011, 'An Australian inquiry into media bias would ignore the bigger crisis facing newspapers in the digital age', *The Age*, accessed December 29, 2012, available: http://www.theage.com.au/opinion/society-and-cul ture/print-is-at-the-root-of-good-news-20110817-1iy1k.html.

Media Watch 2012, 'What's in a name', Australian Broadcasting Corporation, accessed December 23, 2012, available: http://www.abc.net.au/mediawatch/transcripts/ s3458728.htm.

Nash, C., Chubb, P. & Birnbauer, B. 2009, 'Fighting over fires: Climate change and the Victorian bushfires of 2009', paper presented at the Global Dialogue Conference, Aarhus, Denmark, November 3–6.

Nord, L.W. & Strömbäck, J. 2003, 'Making sense of different types of crises: A study of the Swedish media coverage of the terror attacks against the United States and the U.S. attacks in Afghanistan', *Harvard International Journal of Press/Politics*, vol. 8, no. 4, pp. 54–75.

Olausson, U. 2009, 'Global warming—global responsibility? Media frames of collective action and scientific certainty,' *Public Understanding of Science*, vol. 18, no. 4, pp. 421–436.

Pan, Z. & Kosicki, G. 1993, 'Framing analysis: An approach to news discourse', *Political Communication*, vol. 10, pp. 55–75.

Patterson, T. E. 1991, *Out of order*, Random House, New York.

Park, E. R. 1940, 'News as a form of knowledge: A chapter in the sociology of knowledge', *American Journal of Sociology*, vol. 45, no. 5, pp. 669–686.

Park, E. R. 2006 (1940), 'News as a form of knowledge', in S. G. Adam & R. P. Clarke (eds) *Journalism: The democratic craft*, Oxford University Press, Oxford.

Protess, D., Cook, F., EttemaJ., Gordon, M., Donna, L., & Peter, M. 1991, *The journalism of outrage: Investigative reporting and agenda building in America*, Guilford Press, New York.

Reese, S. D. 2007, 'The framing project: A bridging model for media research revisited', *Journal of Communication*, vol. 57, no. 1, pp. 148–154.

Rowe, R., Tibury, F., Rapley, M. & O'Farrell, I. 2004, 'About a year before the breakdown I was having symptoms: Sadness, pathology and the Australian newspaper media', in C. Seale (ed.) *Health and the media*, Blackwell, Oxford, pp. 160–175.

Samios, Z. 2018, 'ABCs: Newspaper circulation suffers across the board with falls as large as 16%', mumbrella.com.au, accessed December 24, 2018, available: https://mum brella.com.au/abcs-newspaper-circulation-suffers-across-the-board-with-falls-as-large-a s-16-535665.

Scheuer, J. 2008, *The big picture: Why democracies need journalistic excellence*, Routledge, New York.

Schlesinger, P. 1990, 'Rethinking the sociology of journalism: Source strategies and the limits of media-centrism', in M. Ferguson (ed.) *Public communication: The new imperatives*, Sage, London, pp. 61–83.

Schlesinger, P. 2009, 'Creativity and the experts: New Labour, think tanks and the policy process', *International Journal of Press/Politics*, vol. 14, no. 1, pp. 3–20.

Schudson, M. 2006, 'The trouble with experts—And why democracies need them', *Theory and Society*, vol. 35, no. 5–6, pp. 491–506.

Schudson, M. 2011, *The sociology of news*, 3rd edn, Norton, New York.

Sigal, L. 1973, *Reporters and officials*, D.C. Heath, Lexington, MA.

Signitzer, B. & Prexl, A. 2007, Communication strategies of 'Greenwash Trackers': How activist groups attempt to hold companies accountable and to promote sustainable development', paper presented at the International Association for Media and Communication Research Conference, Paris, July 23–25.

Solesbury, W. 1976, 'The environmental agenda: An illustration of how situations may become political issues and issues may demand responses from government; or how they may not', *Public Administration*, vol. 54, no. 4, pp. 379–397.

Strömbäck, J. & Nord, L. W. 2006, 'Do politicians lead the tango? A study of the relationship between Swedish journalists and their political sources in the context of election campaigns', *European Journal of Communication*, vol. 21, no. 2, pp. 147–164.

Strömbäck, J., Negrine, R., Hopmann, N. D., Jalali, C., Berganza, R., Seeber, G., Seceleanu, A., Volek, J., Dobek-Ostrowska, B., Mykkanen, J., Belluati, M. & Maier, M. 2013, 'Sourcing the news: Comparing sources use and media framing of the 2009European parliamentary elections', *Journal of Political Marketing*, vol. 12, no. 1. pp. 29–52.

Swain, B. M. & Robertson, J. M. 1995, 'The Washington Post and the Woodward problem', *Newspaper Research Journal*, vol. 16, no. 1, pp. 2–20.

Thompson, J. B. 2005, 'The new visibility', *Theory, Culture and Society*, vol. 22, no. 6, pp. 31–55.

Tiffen, R. 1978, *The news from Southeast Asia: The sociology of news making*, Institute of Southeast Asian Studies, Singapore.

Tiffen, R. 2010, 'The press', in S. Cunningham & G. Turner (eds) *Media and communication in Australia*, Allen & Unwin, Crows Nest, NSW.

Tiffen, R., Jones, K. P., Rowe, D., Aalberg, T., Coen, S., Curran, J., Hayashi, K., Iyengar, S., Mazzoleni, G., Papathanassopoulos, S., Rojas, H. & Soroka, S. 2013, 'Sources in the news', *Journalism Studies*, vol. 15, no. 4, pp. 1–19.

Traquina, N. 2004, 'Theory consolidation in the study of journalism: A comparative analysis of the news coverage of the HIV/AIDS issue in four countries', *Journalism*, vol. 5, no. 1, pp. 97–116.

Trembath, B. 2010, 'ABC chairman criticises media's climate change coverage' (interview with M. Newman), Australian Broadcasting Corporation, accessed December 25, 2012, available: http://www.abc.net.au/pm/content/2010/s2842177.htm.

Tuchman, G. 1978, *Making news: A study in the construction of reality*, Free Press, New York.

Tunstall, J. 1971, *Journalists at work: Specialist correspondents, their news organisations, news sources and competitor-colleagues*, Constable, London.

United Nations Department of Economic and Social Affairs 2017, 'World population prospects: The 2017 revision, key findings and advance tables', UN Working Paper No. ESA/P/WP/248, United Nations, New York.

van Dijk, T. A. 1993, 'Principles of critical discourse analysis', *Discourse & Society*, vol. 4, no. 2, pp. 249–283.

Waisbord, S. 2010, 'Rethinking "development" journalism', in S. Allan (ed.) *The Routledge companion to news and journalism*, Routledge, London, pp. 148–158.

Waisbord, S. 2018, 'Revisited mediated activism', *Sociology Compass*, vol. 12, no. 6, pp. 1–9.

Weiler, H. N. 1983, 'Legitimization, expertise, and participation: Strategies of compensatory legitimation in educational policy', *Comparative Education Review*, vol. 27, pp. 259–277.

Zehr, S. 2000, 'Public representations of scientific uncertainty about global climate change', *Public Understanding of Science*, vol. 9, pp. 85–103.

4 Climate of interpretation: Australia and Bangladesh

The politics of climate change in Australia, one of the largest emitters of greenhouse gases in the world, are highly polarised. This polarisation has arguably led to a period of political instability that has not been witnessed over the past 100 years (Chubb & Bacon 2010, p. 51). In contrast, Bangladesh is a victim of climate change and a poster child for the global debate surrounding it. This South Asian country is unilaterally dealing with environmental hazards that can potentially exacerbate its climatic conditions. The developed economy of Australia is still dependent on coal, and its media debate is swayed by "apocalyptic claims and counter claims" (McNair 2014). Bangladesh's emerging economy is shifting from agriculture to manufacturing, and the country is unique in tackling climate change through adaptations that aim to ensure community sustainability and food security (Johnson 2017). In Australia, the ongoing debate is about an appropriate mitigation policy that would achieve a balance between the economy and the environment. Similarly, in Bangladesh, the prioritising of economic development over environmental concerns is under debate. The focus in Bangladesh, however, is not confined to debating policy (Bhuiyan 2015; Sovacool 2017); it also seeks to act on preventing climatic disasters and aspires to achieve "transformational adaptation" (Huq 2019). Climate change attracts intense journalistic scrutiny in Australia because of the country's continued dependence on coal as its primary energy source and the polarised political positions on this issue. In contrast, climate change does not attract a similar level of media attention in Bangladesh (Nassanga et al. 2017), where the various stakeholders are largely in agreement about climate change policy options, although some differences have recently emerged in various public discussions.

This chapter examines these similarities and differences to explore how the issue of climate change was interpreted in the editorial commentaries of selected newspapers from the two countries in 2009 and 2015. These two years are important because, in the intervening period, a significant shift occurred in the global climate policy regime, that is, a shift from the Common But Differentiated Responsibilities (CBDR) principle to the idea of Intended Nationally Determined Contributions (INDC).

The news media in Bangladesh display a sense of urgency on climate action, whereas the public debate in Australia is in a state of continuous chaos in regard to climate policy alternatives. Perhaps for this reason, critics have referred to the climate change debate in Australia as a "wicked" or diabolical problem (Olausson 2009; Chubb & Bacon 2010). Clearly, Australia and Bangladesh belong to opposite camps in relation to the spectrum of knowledge about the consequences of climate change (Roberts & Parks 2007; Shanahan 2009). A comparison of the discursive constructions of climate change in the two countries' news media would enhance understanding of climate change at the global level by clarifying differences between the hegemonic interpretation of climate change by global agents (such as the UN, IPCC and powerful governments) and non-hegemonic local-level actors. Mike Hulme's (2010) argument concerning global knowledge and IPCC is pertinent here. He asserts: "Rather than seeking a consensual global knowledge that erases difference and allows the most powerful to determine what is 'known' we need to pay greater attention to different ways knowledge comes to be made in different places" (Hulme 2010, cited in Kunelius & Eide 2017, p. 7). In this public understanding of the issue, the news media are a crucial platform (Anderson 1997; Olausson 2009) from which journalists provide information to the public, thus contributing significantly to the construction of public knowledge (Kovach 2006; Patterson 2013). For this reason, an examination of the journalistic portrayal of climate change is of significance.

Considering the need for an inter-disciplinary approach to journalism scholarship (Zelizer 2004; Rupar 2007; Waisbord 2010; Nash 2016), this chapter examines how public knowledge is constituted through newspaper editorial comments. In doing so, it draws on relevant perspectives from the fields of sociology, language studies and cultural studies. In particular, Ulrich Beck's concept of "risk" and Teun van Dijk's as well as Norman Fairclough's discourse analysis are employed to understand the "relations of definition" underpinning journalistic discourse. For Beck (2010), "[r]isks are essentially man-made, incalculable, uninsurable threats and catastrophes which are *anticipated* but often remain invisible and, therefore, depend on how they become defined and contested in 'knowledge'" (p. 261; italics in original). The relations of definition refer to the "matrix of ideas, interest, epistemologies, and different rationality claims (scientific, social, legal, etc.) that compete and contend within the field of risk and ecological interdependency crises" (Cottle 2008, p. 78). The journalistic discourses underpinned by these categories that constitute "relations of definition" ultimately influence what can and should be said about threats and hazards related to climate change by experts, counter experts and the lay public in the production of journalistic knowledge (Cottle 2008).

According to Beck (2006), in the second modernity of the world risk society the "relations of definition" assumed greater importance than the "relations of production" that were used to explain industrial capitalism of the first modernity. It is crucial to identify those who influence the definition of the nature of public threats in the media and the prescription of a course of action to address them, since they clearly command a strategic position with considerable communicative power in the interactions among various climate actors (see also Carragee 1993; Cottle 2010).

The debate around climate change is complex, encompassing looming crises in both scientific and economic contexts. It also raises serious moral and political questions. In analysing the discourses surrounding climate change across the globe, it is important to understand how power relations in the field of communication are exercised and negotiated. To this end, it is pertinent to ask the following questions: How do the editorial commentaries in the selected newspapers respond to this encompassing crisis of climate change? To what extent do these editorials legitimise certain powers and stand over others? Wodak & Fairclough (1997) argue that the power relation between media and politics takes the form of either the domination of media over politicians or the exploitation of media by politicians. Van Dijk (1993, p. 17) defines social power "as a specific relation between social groups or institutions," and asserts that powerful groups exercise their power by utilising resources, including symbolic resources such as access to particular speech acts (commands or directives).

In the present context, it would be pertinent to examine whether the newspapers use editorial commentary to issue commands on the debate surrounding climate change and, if so, how such commands influence the debate. The newspaper editorial discursively functions "as close as possible to being an institutional voice of newspapers" (Hindman 2003, p. 71). An examination of editorial responses to the climate change debate would enable us to analyse the Australian and the Bangladeshi press institutions and their roles and responsibilities. Accordingly, this chapter analyses the role of newspaper editorials in defining climate change in these countries.

Editorial commentaries from the four selected newspapers (*The Australian*, *The Sydney Morning Herald*, *The Daily Star* and the *Prothom Alo*) for the six months leading to the 15th and 21st sessions of the Conference of the Parties (COP15 and COP21) to the United Nations Framework Convention on Climate Change (UNFCC) were examined to see how they interpreted non-binding voluntary emissions reduction as a mitigation strategy rooted in the Copenhagen deliberations and signed during the Paris summit. The chapter also examines the relationship between local legislative debate relates or contributes to the debates at the global summits. These local debates are important because the effectiveness of any agreement signed at an international summit hinges on the passage of national legislation. The purpose of the chapter is to see how the editorials explicitly "take an angle or a spin on the reality" (Ytterstad 2014, p. 2) of climate change issues. This analysis is expected to shed light on the context of journalistic operation, a topic elaborated in Chapters 5 and 6.

Newspaper editorials are the crux of climate change discourses. In the main, editorials are based on the reporting of staff journalists and others, and obtain significant commanding power over and above simple recognition and investigation of the "hierarchy of credibility" (Becker 1967) or news-sanctioned patterns of hierarchical access. Editorial commentaries play a crucial role in advancing various scientific and political viewpoints, leading to different forms of social rationality in the mediated field of climate change (Beck 1992; Cottle 2010; see also van Dijk 1993). Kunelius (2012) asserts that editorials "carve out a conceptual space and argue within it, thus describing key possibilities and potentials that the current debate on climate politics offer ... [They] speak with and comment on the dominant language of politics" (p. 37). Editorials serve their readership constituency; they are interpretative texts, delineating the political views of the newspaper and explicitly criticising or favouring certain positions over others (Roosvall 2017, p. 136).

Although newspaper editorial content has rarely been examined, a few useful studies provide a guide for examining climate change editorial commentaries. Eide and Kunelius (2010), for instance, compared editorial commentaries on the Bali (2007) and Copenhagen (2009) UN climate conferences from 19 countries. They showed that editorial discourses were influenced by "domestication," or the perspective that portrays international climate issues in the context of the respective countries' history and current political and economic interests (Eide & Kunelius 2010, p. 18). In a ten-country comparison of editorial content pertinent to the fifth report of the Intergovernmental Panel on Climate Change (IPCC5), Painter (2017) compared media treatment between developed and developing countries in relation to four themes: disaster, uncertainty, opportunity and risk. Although cautious about drawing any "strong conclusion" consistent with previous studies, he found a high prevalence of the "disaster" theme in all countries. Roosvall (2017) also examined editorials pertinent to the IPCC5 report from 11 countries, including nine high-income and two developing countries. She found that "solidarity was foregrounded in the argumentative discourse of editorial materials published in high income countries during the release of the IPCC report" (p. 146). In other words, wealthy countries' editorial discourses contained strong calls for action on climate justice as a collective responsibility. However, according to Roosvall, these calls were not accompanied by any substantial discussion of appropriate remedies, thus rendering this solidarity discourse ambiguous.

These findings from multi-country comparisons of climate summits and IPCC5 coverage provide a useful framework for examining newspaper editorials in Australia and Bangladesh. One crucial finding was that "domestication" was prominent in the coverage of the climate summits while the treatment of the IPCC reports went beyond the national context. A number of reasons for this difference may be identified. While the IPCC conclusively identified unprecedented effects on the planet of the atmospheric concentration of carbon dioxide as a result of economic and population growth, the climate summits were expected to engage in mitigation commitments, particularly among the "carbon dependent" countries, such as the US and Australia, where carbon-intensive industries and products dominate national economies. As a result, these national economies are significantly affected by any commitment made to the UN Conferences of Parties (COPs), leading media and politicians to call for the protection of national or ethnocentric interests (Schmidt et al. 2013). For this reason, it can be argued that, although the editorials on climate change summits were heavily influenced by "domestication," editorials pertinent to the IPCC5 reports were less so. This in turn suggests that news media accept the notion of solidarity in principle but, when it comes to the implementation of the components of solidarity (climate fund, emissions reduction, etc.), they might have succumbed to other contesting political forces to uphold their "cultural and ideological" (Neverla 2008, p. 9) message in their communication practices.

Exploring the interpretation of climate change

Editorial comments will be analysed using Beck's notion of risk to explore how news media define the risks of climate change in relation to the contestation among various social forces (e.g., politicians, experts, activists and businesses). The analysis will adopt discourse analysis, in particular van Dijk's (1993) notion of persuasive strategies and Fairclough's (1992) idea of intertextuality, both of which have been operationalised through editorial commentaries, the headlines of these commentaries, and editorial subject matter. For Fairclough (1992, p. 270), intertextuality is a form of social practice that "points to how texts are produced in relation to prior texts and restructure existing conventions (genres, discourses) to generate new ones." In relation to the process of generating new genres, Hodges (2015, p. 43) observed:

> [A] text can be thought of as an "objectified unit of discourse" (Gal, 2006: 178) that can be lifted from its originating context (*decontextualized*) and inserted into a new setting where it is *recontextualized* (Bauman & Briggs, 1990). In this way, fragments of discourse from one setting seemingly take on a life of their own as they are turned into texts (*entextualized*) and enter into social circulation. (Italics in original)

Drawing on the persuasive strategies and intertextuality, this chapter proposes the following questions to guide analysis of the selected editorial content: How do editorial commentaries approach the issue of climate change? How are the consequences of various decisions discussed? To what extent are these commentaries different or similar in the two periods? Who is held responsible for climate change policy action or inaction? Which forces have shaped the editorial opinion or gained privileged access to establish a particular opinion or viewpoint?

To address these questions, the headlines and topics of the selected editorial commentaries were coded to identify the main arguments, intertextuality and sources of influence. The topics are crucial aspects of journalistic practice and play a key role in the analysis of climate change journalism. Research has shown that "the topics are usually the best recalled information of a text" (van

Dijk & Kintsch 1983, quoted in van Dijk 1991, p. 73). Such recalled information serves an important discursive function, reflecting "many dimensions of the psychology and sociology of news" (van Dijk 1991, p. 71). Journalistic topics differ from topics in other cultural fields where the issue and topic have a mostly similar meaning. This, however, is not the case for news (Rupar 2007). For example, within the issue of climate change, the news topic could be political contestation over climate policies. The topic summarises pertinent information and performs a very important communication function. In most cases, the information in the editorial is based on reporters' narratives, but other crucial materials can sometimes assist editorial writers to evaluate the arguments of various stakeholders. These materials, which include reports, policy documents and other research papers, can illustrate intertextual features of editorial commentaries through contextualisation and decontextualisation of the text or information (Hodges 2015), as discussed earlier. Intertextuality is central to the process of argumentation, where "the negative evaluations follow from the 'facts'" (van Dijk 1993, p. 264).

To understand other aspects of the discursive potential of the editorials, the chapter will examine lexical styles, word choices for positive or negative evaluations, structural emphasis in headlines and leads, and the incorporation of credible witnesses as an element of intertextuality (Appendix 1). All of these elements contribute to the persuasiveness of editorial comments (van Dijk 1993; see Appendix 2, Tables A2.1 and A2.2). As Nord (2001, p. 133) argues, editorials are official statements of the newspapers; they may criticise or favour a position in the public debate (see also Roosvall 2017). They seek to reach readers who are assumed to agree with the newspaper's position. At the same time, editorials also wish to influence public opinion via official statements or persuasive arguments on climate change. The most prominent feature of an editorial is its headline. The message of the headline is explicit and written in an attractive way. The main intention of the headline is to invite its readership and nationally influential elites (e.g., politicians, business leaders and experts) to accept its argument or judgement on certain issues in the public sphere (Pan & Kosicki 1993). Headlines, as the entry point into the editorial, succinctly summarise some important aspects of the issue and topic. They can also convey the political convictions of the particular news organisation (e.g., climate change supporter or sceptic).

As critics have argued, one of the significant weaknesses in current climate change media research is its focus on global climate change issues at the expense of local environmental issues. To address this imbalance, this chapter addresses climate change in relation to both local and global issues in two countries (Australia and Bangladesh) representing the global North and South and in the context of two key climate change summits (Copenhagen and Paris). To facilitate comparison, each country is first examined separately.

Editorial focus: ETS to RET

In Australia, most of the editorials depicted climate change in the form of "binary oppositions" (Cottle 2013, p. 3) between two contending political forces (i.e., the Labor government and the opposition Coalition in 2009, and vice versa in 2015). This binary depiction was not confined to the national context but was also applied to the international scene. As a result, the newspaper editorials argued that the outcomes of climate policy negotiations at both levels were uncertain. As discussed in Chapter 2, there is a close relationship between the fields of politics and journalism, which might explain this depiction of "binary oppositions." The influence of politics on journalism is strengthened by the fact that the professional culture of the latter strongly upholds conflict as a crucial news value. As a consequence, the editorials in the two Australian newspapers adopted different positions during the summits, arguably to toe their respective political lines.

However, the difference between the two publications was less pronounced on the matter of the INDC and Renewable Energy Target (RET) in 2015 than was the case with the proposed Emissions Reduction Scheme and the CBDR in 2009. In 2009, the discursive battle was heavily influenced by contestation between political forces at national and international levels. Although there was consensus between Prime Minister Kevin Rudd and opposition leader Malcolm Turnbull on reaching an agreement over the proposed Emissions Trading Scheme (ETS), according to one editorial, the internal disagreement within the opposition Liberal–National Coalition was pushing Australia "[o]n the road to oblivion" (*The Australian,* November 11, 2009). This article identified the ramifications of deep ideological divisions over mitigation policy as creating risk and uncertainty for both climate policy and the conservative forces in Australia, who were heavily influenced by "industry lobbying, political culture, global deadlock and a resurgent scepticism." According to this article, the uncertainty (Painter 2013, 2017) stemmed from the "scare tactics of climate catastrophist" and fuelled the doubts of the climate deniers (such as then opposition Senate leader Nick Minchin). It explicitly suggested that the aggressive strategies supported by Prime Minister Rudd were responsible for the deep polarisation over climate policy in Australia.

Leading up to the Paris Summit, the editorial coverage overall demonstrated implicit acceptance of various climate policy measures, albeit with some reservations at both national and international levels. Following van Dijk (1988), the editorial headlines were analysed to identify the "relations between words" that described the roles and relationships between news actors. Most of the headlines were non-specific in terms of attribution of responsibilities. A small number, however, identified the political parties (Liberal and Labor), the Prime Minister and environmental extremists as responsible for a deadlock in climate negotiations within and outside the country. Here, the action of the environmental activists was assessed negatively in line with the newspaper's sceptical stance on climate change. The definition of prevailing circumstances in climate change is dominated by elite discourses (Beck 2010), notably that of politicians. While most of the editorial headlines were non-specific or neutral in relation to the commentary's intention, this was clearly established through their topicalisation. Topicalisation involves selecting what to prefer in the topic position to influence readers' perceptions in a particular direction (Huckin 1997; McGregor 2003). A neutral headline could be understood as the newspaper providing a public voice and seeking to reach its readership (Meyer 2001; Rupar 2007). This neutrality offers greater scope for interpretation by not identifying or attributing responsibilities, instead alluding to the responsible agents for mitigation action or inaction.

The following headline clearly adopts a neutral stance: "Taking a sensible route to Paris" (*The Australian*, November 28, 2015). A close analysis of the editorial, however, shows that it goes beyond informative discourse and applies persuasive strategies. This simplistic headline appears to invite its readers—and everyone else—to be reasonable in their approach to the policy commitments made at the Paris conference. And why would anyone question a sensible approach to climate mitigation? The headline, however, implies that some agents in the global negotiations may not be sensible. The rhetorical assertions in the editorial gain credibility through the use of intertextual strategies, such as citing economic modelling data to demonstrate that the opposition Labor Party's policy would be costlier than that of the Coalition government. It praises the government's "Direct Action" model without mentioning either the cost involved or any of the criticisms of the model that had emerged in public discussion (Lubcke 2013). It also links the proposed renewable policy to the previous Gillard government's carbon tax. Finally, the editorial refers to one of the newspaper's own business reporters to argue that "renewable is not reliable." The use of these sources can be understood as part of a persuasive strategy involving an intertextual connection. In essence, the editorial uses the notions of "credible witness" and "symbolic resources" to sway readers in favour of its argument and to introduce the notion that "renewable is not reliable" into the social circulation of meaning (van Dijk 1993; Hodges 2015). In other words, public acceptance of the sceptical position on climate change is sought by associating it with economic rationality (Bulkeley 2001; Christodoulou 2019).

Editorial topicalisation

The topicalisation of climate change was examined by reading the headline, introduction or first paragraph, and identifying the trigger or event that led to the writing of the editorial. Intertextuality was also examined to understand how the attribution of responsibility for any action or inaction was justified and which forces or groups were given privileged access. The analysis identified various elements of the discourse structures discussed earlier (argumentation, dominant and dominated forces, responsible agents, etc.). Almost two-thirds of the articles were triggered by national climate politics, and the rest by international climate politics related to COP15 and COP21. Consistent with some previous investigations (Chubb & Bacon 2010; Bacon & Nash 2010), this study found that *The Australian* argued against the proposed ETS, highlighting the necessity for global action to cut the cost of "climate pain" (e.g., "Only global action can cut cost of climate pain," *The Australian*, August 12, 2009). Understandably, the newspaper presented a negative evaluation of the climate change policy using metrics that prioritised economic risks over climate risks and normalising the state of neglect of measures to address climate change. This has been described as the "business as usual approach" (Goodman 2016).

Most of the editorials in 2009 were triggered by the proposed ETS. In 2015, by contrast, the RET was one of the main triggers in the public debate. The political antagonism was reflected in the differing positions of the two newspapers on climate policy. Although focus here is not on the difference between the editorial positions of the two newspapers, for the sake of overall generalisation the similarities and differences were taken into consideration. As one aspect to the debate, the editorials broadly argued that some proposed climate policies (i.e., the ETS or RET) were a "costly challenge" to the Australian economy, and that carbon pricing required more rational and rigorous debate. In 2009, the focus of this negative evaluation was on the Labor government's proposed ETS; in 2015, the focus was again on Labor's RET policy, although the party was in opposition this time around. In other words, the Labor Party's climate policies attracted consistent negative evaluation in the climate debate irrespective of whether it was in power or opposition. In fact, the editorials were not influenced by the political party as such, but by the policy. For example, an important focus of the editorial topics in both newspapers in 2009 was the "crack in Coalition," that is, the (then in opposition) Liberal–National Coalition. In one editorial, *The Australian* stated that the "Liberals need to get back to the fights they can win" (December 1, 2009). It referred to climate sceptic Liberal leader Tony Abbott's change of mind in a bid to unseat the then Liberal leader Malcom Turnbull. *The Sydney Morning Herald* similarly urged that it was "Time to end climate comedy" (August 10, 2009), and expressed its concerns about the Coalition's lackadaisical approach to the proposed climate policy—the Carbon Pollution Reduction Scheme (CPRS).

In this discursive battle, Beck's (1992) idea of "different rationality claim" was played out to establish binary positions. On one side, concern was raised about the potential economic consequences of carbon pricing; on the other, concerns about climate, influenced by various scientific underpinnings, were identified as the reasons for the Australian government's proposed carbon policy. Throughout this period (July–December 2009), *The Australian* proclaimed "Climate scheme needs better scrutiny" (September 26, 2009), proposing that the Australian government's climate policy could have a profound impact on the Australian economy. This editorial cited former Prime Minister

Kevin Rudd's statement that the proposed ETS was not a "political slap and tickle." It referred to another Australian political debate on the introduction of a Goods and Services Tax (GST) in the mid 1990s to compare and recontextualise (Hodges 2015) the climate policy debate, arguing: "It [ETS] is a serious legislation that could potentially have a greater impact on productivity, capital flows and jobs than the GST, which was subjected to intense scrutiny and almost cost the Howard government office 11 years ago."

Several sources of evidence, including a survey by the Queensland Chamber of Commerce and Industry (QCCI), were used to demonstrate the desperation of the business community to know more about the consequences of the introduction of a carbon reduction scheme on those industries that were isolated from the "warming or cooling" debate. The editorial essentially made the government—the proponent of the scheme—responsible for not providing adequate information for the public to engage in policy debates. This is one of the ways in which the negative evaluation of climate policy was rationalised by making an explicit intertextual connection with a previous national policy debate, by prioritising business concerns, and by identifying the agents who would be responsible for a negative impact on the nation's economy. It further stated that the environmental argument should be economically rational. The proponents of the economic rationality argument essentially attempted to normalise the debate around climate policy via a negative assessment of activists from non-governmental organisations (NGOs) at the Copenhagen Summit and the actions of the Australian Greens in the national climate debate: "After demanding the demise of the nation's biggest export industry, coal, the Greens' condemnation of the $50billion Gorgon liquefied natural gas project off the West Australian coast again showed how far removed the party was from mainstream Australia" ("Green debate must not alienate voters," *The Australian*, August 27, 2009). The editorial referred to QCCI Director David Goodwin, who believed that "an emissions trading scheme that encouraged such industries moving offshore would be counter-productive by contributing to carbon leakage in the global economy" (*The Australian*, August 27, 2009). The proponents of the other side of the debate were, however, frustrated by slow progress in climate policy. *The Sydney Morning Herald* made the then Prime Minister Kevin Rudd's government responsible for its failure and later labelled the Prime Minister a "Rudderless leader" (July 30, 2009) who induced "paralysis of hope, a loss in faith in our ability as a nation" to resolve a complex problem. *The Sydney Morning Herald* editorial headlines explicitly identified Kevin Rudd as the agent of inaction in advancing climate policy to address emissions reduction in Australia. However, the editorials in this newspaper mainly focused on the political failure of the Prime Minister as a leader and provided no further evidence on the consequences of irresponsible leadership in addressing the Australian climate debacle. During the period ahead of the Paris summit in 2015, in the context of the ongoing political issues surrounding climate policy, the newspaper welcomed the return of the "first martyr" of climate politics—Malcolm

Turnbull—in efforts to tackle global warming. It argued that "Turnbull's agility must strengthen Direct Action" (*The Sydney Morning Herald*, October 16, 2015). The newspaper mainly engaged in political analysis instead of scrutinising or evaluating the Coalition's "Direct Action" policy. Interestingly, the newspaper was critical of the Coalition during the Copenhagen summit, but offered a privileged space to the Australian government position by directly quoting the then Prime Minister Malcolm Turnbull, who used the word "agile" many times to demonstrate his determination to act.

From the above discussion, it can be deduced that, while *The Australian* relied on business actors to mobilise support for the argument underpinned by economic interests (Anshelm & Hultman 2014), *The Sydney Morning Herald* tended to rely on political analysis, raising questions about the accountability of the then Labor leader and former Prime Minister Kevin Rudd for future climate risks. Following several critics (van Dijk 1993; Wodak & Fairclough 1997; Hodges 2015), it can be argued that, in editorial terms, *The Australian* adopted a persuasive strategy involving intertextual connection while *The Sydney Morning Herald* did not necessarily follow a similar strategy. This is an important consideration because *The Australian* has more access to the Australian people, in terms of circulation and the number of articles pertinent to climate change, than *The Sydney Morning Herald*, despite the fact that the latter is published in the largest city, Sydney (Farbotko 2005).

Innovation vs subsidy

As previously discussed, in 2015, *The Australian* continued its denigration of climate policy. One of its editorials, titled "Labor's loony turn on the renewable energy target" (*The Australian*, July 23, 2015), labelled the proposed emissions reduction target as "seat-of-the-pants policy making." It described the Labor Party's RET as political opportunism designed to reach Australian Greens voters rather than a "climate action" as the Labor Party had claimed. Similar to the proposed ETS in 2009, albeit in a different manner, *The Australian* evaluated the renewable energy policy on two fronts—it would not be cost efficient and would not be a reliable source of energy, thus jeopardising our energy security.

The renewable energy debate as a national policy issue crept up in the international policy debate during the Paris Conference. On the eve of the conference, *The Australian* ran a strong editorial that encapsulated the newspaper's position on some crucial climate policy issues, such as energy policy and climate aid. The editorial argued that "Climate change demands innovation, not subsidy" (*The Australian*, December 2, 2015), claiming that the impetus to innovate had at times been misrepresented by environmentalists as a manifesto for inaction. "It is true that the cost of wind and solar are falling rapidly and both can now be competitive at low levels of grid penetration," it quoted Dr Jonathan Symons of Macquarie University as saying. But, the editorial continued, because of associated costs, Australia should not act alone ahead of its northern hemisphere counterparts. The innovation versus subsidy argument was not confined to the national renewable energy sector, but also extended to the matter of foreign aid in the same editorial:

> There is a pseudo controversy over climate mitigation and foreign aid. In Paris, Mr Turnbull announced a five-year diversion of at least $1 billion from the foreign aid budget to climate mitigation projects in the Pacific. Labor's complaints ring hollow. Only last month Bill Shorten toured the Pacific (remember the prophesied climate refugees?)
>
> (*The Australian*, December 2, 2015)

Overall, the editorial showed strong opposition to any kind of assistance, raising questions about its efficacy. Climate aid had encountered similar difficulties in 2009. During the lead-up to the Copenhagen Summit, the editorials evaluated the role of Australia in the conference positively but, at the end of COP15, *The Australian* editorials identified "environmental extremist" views as responsible for the summit's failure. It asked: "Is Copenhagen just an exercise in wealth redistribution?" (*The Australian*, December 17, 2009). The newspaper observed that the summit stalled because of conflicting agendas between developed and developing countries. Another editorial ("The parallel universe with a life of its own," *The Australian*, December 17, 2009) stated:

> THE shemozzle in the Danish capital has done little to encourage voters' trust in the global efforts to get to grips with climate change. Even as leaders focused overnight on tax measures to try to achieve a breakthrough at the Copenhagen summit, the conflicting agendas continued to cloud its central task. Derailed by efforts to end world poverty and redistribute wealth under the guise of helping the developing world adapt to climate change, the meeting has appeared unfocused and even dysfunctional in recent days. The negotiating process has been distorted by NGOs and developing nations with a cargo cult mentality who see climate change as a way to attract aid dollars from the West. (Emphasis in original)

In 2009, the newspaper described the negotiation process of both the NGOs and developing countries as reflecting a "cargo cult mentality," in which negotiation was seen as a way to attract aid dollars from the West. In 2015, the climate debate was underpinned by mitigation measures, particularly with the introduction of INDC, and the newspaper's editorial writers labelled climate mitigation and foreign aid as a "pseudo controversy." Such labelling of the needs of vulnerable developing countries can be understood as "exclusionary and stereotypical" (Said 1978; Harvey 1993, cited in Farbotko 2005, p. 4). It is important to note that, although *The Sydney Morning Herald* was strongly in favour of global climate action, it did not explicitly address the imperatives of climate assistance to the affected developing countries. As a result, *The Australian*, through its negative evaluation of foreign aid, sustained and recontextualised its position against climate funding by raising questions about the process of "mission innovation" in the global climate negotiations. Australia had promised to double its clean energy research and development as part of the 20-nation project known as Mission Innovation. Although *The Australian* was less inclined to support the notion of solidarity in global climate action, *The Sydney Morning Herald*'s mission was to reveal underlying factors, such as business interests, that affected the progress of climate policy. The newspaper even supported "healthy scepticism" as a good sign, arguing: "Scepticism, honest doubt about the truth of a particular fact or theory, can be a healthy thing. But the term becomes debased when used as a euphemism for self-interested denial, or wilful ignorance" ("The abuse of skepticism," November 9, 2009). Referring to reports produced by the International Consortium of Investigative Reporters, it found:

> In Australia, 120 companies with significant greenhouse emissions use about 80 lobbying firms, and their own in-house PR people. There are more than 2800 climate lobbyists in the US. It works, too. Rudd made huge concessions to big polluters before arriving at the present watered-down form of his Emissions Trading Scheme.

Unlike previous editorials, this article successfully (through intertextuality) conjured up the image of Rudd as an inefficient politician ("Rudderless leader") by using evidence of anti-climate lobbying to show that he had failed to tackle climate change, thus failing Australia in this regard. This is an example of a new entextualised position, in which fragments of discourse from one setting seemingly take on a life of their own as they are turned into texts and enter into social circulation. Essentially, the "Rudderless leader" emerged in 2008 in the context of the first climate change election in the world, during which Rudd was depicted as a climate saviour.

Overall, *The Australian* maintained its sustained negative evaluation of climate policy in both national and international contexts. Its counterpart *The Sydney Morning Herald* addressed the issues on two fronts: the Coalition and the Rudd government. This newspaper was inclined to assess the government positively when Malcolm Turnbull was installed as Prime Minister in 2015. *The Australian* successfully recontextualised its position in 2015 (in the midst of the climate mitigation debate) by connecting it to its economic rationalist position. The intertextual connection in *The Australian* was slightly more rigorous than that in *The Sydney Morning Herald*. The topical focus on national climate change policy in 2015 was contested but not to the extent it had been in 2009. Perhaps, by then, the contestation over climate policy had been normalised.

> Five years is a very short time when it comes to measuring global warming. It is an age, however, in terms of assessing the politics of climate change. The political momentum for concerted global action looks much better today

than it did five years ago when the Labor government of Julia Gillard started preparing for the ill-fated carbon tax. In recent months many nations have pledged to reduce emissions further than expected. Technology is evolving rapidly towards the goal of cleaner emissions ("Climate for Change: Why Paris matters so much?" *The Sydney Morning Herald*, November 24, 2015).

The shift in the argument of failed Copenhagen (Eide & Kunelius 2010) in 2009 to "Paris matters" in 2015 can indicate a need to move beyond the limits of "methodological nationalism" (Beck 2015, p. 76), in which almost every issue is framed in reference to a nation state (see also Olausson 2009; Eide et al. 2010), thus creating a narrow perspective on the issue of climate change.

In 2015, the coverage in *The Australian* can be understood as a "business as usual" approach underpinned by economic rationalist logic, albeit from a slightly different perspective compared to the usual economic rationale. This difference relates to the existing political environment in which, as one editorial in *The Australian* put it, "no major party will adopt a climate refusenik position" (*The Australian*, November 28, 2015). *The Sydney Morning Herald's* reference to a changing "community climate" can be understood as the "politics of possibility" (Beck 2015) emerging from a "new realism" (Eide & Kunelius 2010, p. 42), underpinned by the shift from disagreement at the 2009 Copenhagen summit to a historic agreement at the 2015 Paris meeting. This, however, begs the question: Is it an emancipatory side effect of global climate risk? This question arises in particular from *The Australian's* coverage, which used persuasive strategies through intertextuality (e.g., contextualisation and recontextualisation) to consider only the previous policy implications (e.g., GST) and ramifications for the economy (i.e., cost challenges, job losses) while remaining cognisant of the fact that no one can be a "climate refusenik" and that "mission innovation" is needed to tackle the climate problem.

Bangladesh

Bangladesh, the "nature laboratory" (Inman 2009; Das 2015) and poster child for climate change, is located in the world's largest delta. As an emerging economy, Bangladesh is one of the most vulnerable countries to climate change and the associated rise in sea level because so much of it is flat deltaic land. It is increasingly vulnerable to climate-induced disasters including storm surges, cyclones, flash floods and increasing salinity (IPCC 2007, 2014; Dastagir 2015). According to the Grantham Research Institute on Climate Change and Environment (2018), this lower middle-income country experiences an annual loss of Gross Domestic Product equivalent to 2.568 per cent. Bangladesh is continuously dealing with both external and internal environmental factors. While global climate change is projected to result in increased temperatures and other climate-induced hazards, Bangladesh also faces internal issues, such as a large population and poverty, which exacerbate the country's environmental challenges. There are 1,015 people in every square kilometre of this small South Asian nation, and about 50 million people in Bangladesh live in poverty. While Bangladesh is a victim of climate change, questions have been raised about its capacity to fully implement existing climate measures in the current political and economic environment (Bhuiyan 2015; Sovacool 2017; Rahman 2018). However, Bangladesh has been at the forefront of efforts to tackle climate change. It ratified the Kyoto Protocol in the 1990s and, more recently, entered into the historic Paris Agreement by committing to INDC to reduce its own greenhouse gas emissions. Local scientists started to assess the impact of global warming in this low-lying land almost three decades ago; the impacts they identified included destructive cyclones, heavier rainfalls and rising sea level (Hanlon et al. 2016). The Ministry of Environment and Forests (MoEF) is a key government agency responsible for climate change. The highest-level plans to address the domestic impacts of climate change are the National Adaptation Programme of Action (NAPA) and the Bangladesh Climate Change Strategy and Action Plan (BCCSAP), both published by the MoEF. The country has also developed a number of climate laws including the Disaster Management Act 2012, the Sustainable and Renewable Energy Development Authority Act 2012, the Climate Change Trust Fund Act 2009 and the Bangladesh Energy Regulatory Commission Act 2003.

The news media coverage in Bangladesh was dominated by a victim perspective. In the national political context, discussion of climate change was heavily preoccupied with future environmental hazards that would exacerbate the climatic problems of the deltaic plain. Media reporting of climate change was mainly triggered by various environmental conferences, including the COPs (Nassanga et al. 2017), as well as the implications for local development projects on ecological protection. Media reporting on the latter tended to be in the form of an environment versus development debate. The headlines of the Bangladeshi editorial commentaries pertinent to the Copenhagen and Paris conferences that are examined in this chapter were explicit and interrogative, although the connections between the local climatic issues and broader global climate change debate and Bangladesh's role in that debate were mostly invisible in 2009 and 2015. Local issues that attracted topical focus included, in 2009, the proposed Tipaimukh Dam across the border in India that would heavily impact on the overall environment of Bangladesh and, in 2015, the proposal to build Rampal Power Plant in the country's southwest. The editorial commentaries pertinent to these topics described potential hazards arising from them, but the issue of climate change was tangentially related to as one of the challenges to sustainable ecology and biodiversity. One of the articles, titled "Environment: How we understand climate change" (*Prothom Alo,* December 12, 2009), directly linked Tipaimukh Dam with potential climatic risks to the environment of Bangladesh. It described some scenarios as "dangerous climate catastrophes" (Beck 2006) and emphasised the importance of bringing them to the attention of readers.

Underpinned by a deep-seated news value of deviance and negativity (Cottle 2010), the editorials also attempted to envisage invisible hazards (e.g., "the Sundarbans under threat from the Rampal") which had the potential to gain more traction in the journalistic discursive practice. Unlike in Australia, however, where the environmental coverage was directly linked to a "binary opposition" between government and opposition parties, Bangladeshi media did not explicitly identify these issues as a contestation between the two strong political forces. Instead, the debates were predominantly fronted by experts and civil society groups, who endeavoured to represent various political interests and reach the policy-makers' ears.

From Tipaimukh Dam to Rampal Power Plant

The following topical analysis focuses on the discursive elements and strategies of argumentation, rhetorical structure and intertextuality in the editorial representations from Bangladesh. Overall, the editorial commentaries drew attention to potential future climatic catastrophes in the context of the environmental degradation in the southwestern part of the country resulting from India's construction of Farakka Barrage on the river Ganges. A case in point here is a five-part series of opinion pieces in the *Prothom Alo* by a former bureaucrat-cum-expert, Akbar Ali Khan. This series, published during September 2009, gave an account of water politics in South Asia and outlined the unprecedented potential consequences of constructing a dam (Tipaimukh) in such a densely populated region of the world. It noted a lack of research on the side effects of the proposed dam, and claimed it would bring anthropogenic catastrophe to the region, particularly in lower riparian Bangladesh. The articles referred to a number of Indian sources that expressed ambivalence about the necessity for and consequences of such a mega-dam, arguing that the "magnitude of risk" would be quite high. In another editorial, titled "India and Bangladesh: Quest for eternal and permanent national interest and tip of the iceberg" (*Prothom Alo*, September 7, 2009), the newspaper expressed concern about the dominance of its upper riparian neighbour:

> We have to let the world including the Indian people to know the problems with this project. For that, we need a thorough world-class research led by Bangladeshi and international researchers. The most crucial matter is that the Bangladesh government should be woken up from sleep. If the government fails to realise the severity of the issue, then none of the parties would recognise the possibility of the serious environmental degradation.

The commentary demonstrated that there had been inaction from the then government, which prompted the writer to call for more research by local and overseas experts who could be seen as "legitimate, credible definers of reality" (Allan et al. 1999) in the media politics of environmental risks. The article clearly showed Bangladesh's vulnerability in terms of inadequate scientific knowledge about the proposed dam as well as a lack of political will to challenge its powerful neighbour. The imperatives of expert's opinion were also expressed in an editorial in *The Daily Star*, which directly challenged the official position: "Why cloud the atmospherics for better ties?" (July 4, 2009). This newspaper column highlighted India's position through statements by its High Commissioner to Bangladesh, Pinak Ranjan, who completely denied the possibility of negative consequences from the proposed Tipaimikh dam, sparking a strong reproach from both Indian and Bangladeshi experts. The High Commissioner was stated as saying that: "There is absolutely no evidence that dams cause earthquakes." The editorial directly refuted this position with reference to the burgeoning body of scientific evidence, such as the Sichuan earthquake in May 2008, which killed 80,000 people instantly. Describing the earthquake associated with the dam as "Reservoir-induced Seismicity" (RIS), Indian scientist Dr V. P. Jauhari argued that: "There is positive correlation between the height of the water column in the reservoirs and the seismicity induced" (*The Daily Star*, July 4, 2009).

The topical focus in the anti-dam position alludes to the political tension between the two neighbouring countries, implicitly associating the construction of the proposed dam in India with a "creeping catastrophe" (Beck 1996, as cited in Allan 2002, p. 100) in Bangladesh. This claim of looming climatic disaster was not completely unjustified given the country's experience with the Farakka Barrage on the Ganges just a few kilometres from its Western border. This barrage, constructed by India in the 1970s, had already left a devastating ecological footprint on the flow of water in several tributaries of the river Padma and other downstream rivers, and had heavily impacted on agricultural production and bio-diversity in the southwestern part of Bangladesh. In the process of intertextuality, the recontextualisation (Hodges 2015) of the Farakka experience was useful. However, this reference was not sufficient to entextualise or offer a fresh meaning to the issue of the dam. To highlight the potential for a public outcry, the editorial anticipated the ecological consequences of the proposed dam in the country's Eastern Sylhet region, which currently enjoys the benefits of upper stream water flows from India. The argument relied heavily on a negative evaluation of the dam based on strong intertextual evidence drawn from experts and historical experiences as a lower riparian country. Through the use of expert evidence and historical experiences, the editorial sought to influence the "relations of definition" by challenging India's political domination in the region.

Interestingly, in 2015, the focus of environmental editorial commentaries shifted from political domination of the neighbouring country to the Bangladesh government. In this period, the editorial commentaries directly challenged the government's proposed coal-fired Rampal Power Plant, characterising it as potentially damaging to the "mighty mangrove forest" of the Sundarbans. This editorial challenge emerged in mid-2015 and continued sporadically in the period leading up to the climate conference in Paris. Special editorial opinion columns criticised the government decision because of the "risks it posed to the critical ecological area." An article titled

"Sundarbans under threat" (*The Daily Star*, July 25, 2015) presented evidence to question the financial acceptability and environmental logicality of this project. Referring to Bank Track, a coalition of organisations tracking the financial sector across the world, the article asserted that there were "serious deficiencies in the project design, planning and implementation." While the editorial writers adopted a unilateral stance against the proposed Tipaimukh dam in 2009, they provided space for contending views on the proposed power plant at Rampal in 2015. In the *Prothom Alo,* an article titled "'Rampal Politics' over coal" stated:

> The prospect of Bangladesh becoming a middle-income country is increasingly bright. One of the crucial aspects of expediting economic growth is electricity. As Lenin once said, Marxism +Electricity = Revolution. If there is no electricity, there would be no light. We must increase electricity production and transmission. But how?
>
> (*Prothom Alo*, July 24, 2015)

Referring to the CIA *World Factbook*, the author justified the demand for increased electricity generation on the grounds that, in terms of per capita electricity consumption, Bangladesh's average of 28 kilowatt-hours (kWh) was far behind that of its neighbours Sri Lanka (52 kWh) and India (90 kWh), while Australians consumed on average 1,114 kWh of electricity annually. One may argue that a positive assessment of the plant would be influenced by the conventional market economy, which prioritises development over environmental protection, and a negative assessment would favour the ecological economy, which adopts the view of "economic incommensurability" that values the benefits for future generations and species over the monetary value of environmental resources (Guha & Martinez Alier 1997, p. 23).

Although the influence of conventional economic theory is evident in the debate, this influence cannot be taken for granted. However, the notion of ecological economy was not clear in this debate. In the case of both Tipaimukh and Rampal, experts and other political forces urged the government to work against or modify the plan to lessen any potential ecological damage. The opinion columns were not completely against the Rampal Power Plant but many columnists pleaded for the proposed site to be relocated so that it did not damage the mangrove Sundarbans. These columns reminded readers that this UNESCO world heritage site was public property and should be protected for future generations. The debate over the Rampal Power Plant resurfaced during the Paris conference, as environmental activists from both sides of the international border (i.e., Bangladesh and India) attempted to bring this issue to world attention but were unable to engender enough interest. Here, it is pertinent to mention that the protests against both the Tipaimukh Dam and the Rampal Power plant in 2009 and 2015, respectively, received cross-border attention. Critics described it as "environmental resistance beyond border" (Islam & Islam 2016, p. 2). In building the pro-environmental position, both newspapers extensively used expert opinions, research papers and other reports to challenge the decision-making authorities. The topical focus in 2009 on India's unilateral decision to construct the Tipaimukh Dam demonstrated the "indeterminacy" or uncertainty associated with the potential climatic hazards from the dam. The Rampal power plant case in 2015 was different because it was not a unilateral decision but a joint venture of the Bangladesh–India Friendship Power Company. Understandably, the news media's pro-environmental concerns did not last long, giving way to the developmental priority in favour of coal-fired electricity generation. The governments of both India and Bangladesh went ahead with the proposed development projects without paying any heed to public concerns (Islam & Islam 2016; Lemly 2017).

Both debates were dominated by experts, civil society organisations, and some political organisations, but the media coverage was characterised by the notable absence of any leaders from the ruling political party, Awami League, or its important opposition, the Bangladesh Nationalist Party. These absences can be explained by the notion of the "politics of invisibility" (Beck 2016, p. 99) that engender a new landscape of "relations of definitions." In this landscape, actors can seek advantage not only through visibility but also through inbuilt or nurtured invisibility. In environmental issues, media invisibility can be inbuilt or taken-for-granted; for example, the invisibility of catastrophic future risks. Manufactured or nurtured invisibility is achieved when actors deliberately remain invisible in the public discussion. In this particular case, powerful government leaders ensured their invisibility by not participating in the public debate over Rampal. This invisibility enabled them to ignore public concerns about the proposed plant and avoid scrutiny.

Domestic consequences of global crisis

The headlines of editorial articles pertinent to COP15 and COP21 raised questions about the climate summit negotiation process: "Can Copenhagen deliver hope?" (*The Daily Star*, December 19, 2009) and "Gambling climate" (*Prothom Alo*, December 7, 2009). Over time, this interrogation became sharper and more explicit: "Will Bangladesh be compensated for the loss and damage?" (*The Daily Star*, November 29, 2015). The headlines explicitly showed the risk and responsibility divide in the debate over climate change (Billett 2010). Some of the headlines clearly outlined the international community's neglect of the country's climate challenges: "Everybody is talking about Bangladesh, but who is listening?" (*Prothom Alo*, December 9, 2009) and "A call for climate justice" (*The Daily Star*, December 5, 2009). At times, the sentiment might seem to have been normatively enforced (Smith & West 1996) through the headlines. For example: "Paris Treaty is treacherous to the world" (*Prothom Alo*, December 26, 2015) and "Climate disaster: The world will be saved if Bangladesh is saved" (*Prothom Alo*, December 4, 2009). The comparative analysis shows that interrogative headlines were more prevalent during the Copenhagen conference than the Paris meeting. In 2009, the government of Bangladesh expected outcomes from the summit and attempted to pursue its position by directly raising various questions. These questions were reflected in the editorial headlines. In 2015, it seems that the editorials only responded to the questions that were raised, although the editorial writers continued to argue that the world would face dire consequences if they failed to act in the universal interest of saving the planet.

Overall, it can be suggested that the editorial headlines demonstrate the continued discursive intensification of climate risks for Bangladesh because of the lack of progress at the COPs. The use of words such as "frustration," "treacherous," "disaster" and "compensation" captures the atmosphere of despair and gloom surrounding the climate conferences. The call for global solidarity continued even in 2015 when the coverage was significantly reduced. The pattern of editorial headlines can also be understood as a process of normalisation; that is, normalisation of the existence of gloom or uncertainty. As well, they can be seen to encapsulate the demand for "climate justice" from global communities by using Bangladesh as an example to illustrate the consequences. The editorial commentaries were diverse. They reaffirmed the risk-responsibility divide (Billett 2010) in the climate change debate that identified the country as at high risk of climate-induced disasters because of past industrial developments in Western countries. However, this divide was characterised by national interest (Olausson 2014, p. 711) that engendered serious "domestication," and was more prevalent in *Prothom Alo* than in its English counterpart. The editorials in the former were proactive in sharpening the focus on empirical evidence of global crises during both COP15 and COP21, and strongly argued for climate justice from developed nations. The editorials in *The Daily Star* were also supportive of this demand, but expressed the support in a more restrained manner:

> It's a position the developed world must not ignore. Prime Minister Sheikh Hasina has made a plea at the Copenhagen Summit for compensatory grants ... The theme of climate justice is precisely what came through Sheikh Hasina's laying out of a position for Bangladesh and other affected nations.
>
> ("Bangladesh leader's call in Copenhagen," *The Daily Star*, December 18, 2009)

Referring to the then UN Secretary General, Ban-Ki Moon, another editorial strongly supported the "US-brokered deal" as a significant step forward in achieving the goal of keeping the rise in global temperature to 2 degrees Celsius (*The Daily Star*, December 21, 2009). This editorial implicitly rejected the notion of "abject" or "miserable failure" of the Copenhagen conference "for the simple reason that they highlighted the massive degree of concern all nations have about the effects of climate change." In 2015, the editorials were oriented towards "Preparing for the future: We must not wait till it's too late" (*The Daily Star*, November 21, 2015). Bangladesh was praised for taking "commendable steps" to mitigate the effects of climate change. The editorial demonstrated a strong conviction that global action to decrease emissions was needed if the countries were to avoid the "bleak probabilities" resulting from catastrophic climate change. Both newspapers made explicit empirical connections between climate change and local communities, and strongly argued for adopting the World Health Organisation's recommendations for appropriate adaptation measures, such as constructing dikes and climate-resilient health centres in coastal zones. Often, however, the editorial arguments, which were sometimes based on in-house reporting (see Chapter 6), lacked connection with the global climatic ramifications. This can be explained as "introverted domestication" where a "climate issue [is] deprived of its global character when constructed as an entirely domestic concern" (Olausson 2014, p. 715).

Solution scepticism

Unlike the editorial articles of 2009, those of 2015 argued that Bangladesh should assume some responsibility in tackling its own climate change-related issues. During this period, the editorial commentaries in both newspapers continued to focus on the possibility of keeping the "pledges" made by developed nations (Huq 2015) to slow down climate change by adhering to the long-term temperature rise goal of 1.5 degrees Celsius. During both study periods, the call for a climate fund did not receive unilateral support, perhaps because of the country's poor record in local environmental protection, which had been reported in studies that examined environmental issues other than those related to vulnerability (Bhuiyan 2015; Sovacool 2017). A number of editorial articles published in 2009 and 2015 in both newspapers presented critical domestication views; that is, they scrutinised both the government and international policy goals (e.g., UNDP) that prescribed certain development structures that were not conducive to climatic or ecological protection. One of these ("Mirroring Time: World climate vs. Bangladesh environment," *Prothom Alo,* December 24, 2009) referred to the idea of a climate fund for Bangladesh and questioned the efficacy of reducing carbon emissions through financial aid. In other words, it questioned whether the relentless emission of carbon could be compensated for financially. It asked:

> Which activities engender pecuniary benefit for certain sections of different regions? The important question here is: Who would ensure that the way this climate fund was utilised would benefit the climate-affected people of Bangladesh? The reality was that the ordinary citizens of Bangladesh, who had traditionally been in the forefront of climatic disaster, mostly tackle these challenges by themselves.

Similarly, an editorial commentary titled "Promises or rhetoric: Climate change and SDG" (Sustainable Development Goals) (*The Daily Star,* December 5, 2015) raised questions about the intentions of the world political leaders across who signed the declaration on the SDG:

If we look at Bangladesh, despite endorsing SDGs and remaining active in climate change negotiations, the country is adding to the areas of unsustainable development (USD). USD in Bangladesh includes: several flood control and irrigation projects that are killing rivers; the Rampal coal fired power plant that will possibly destroy the Sundarbans; the construction boom through encroachment on agri lands, wetlands and canals ... grabbing of hills, rivers and open space in the name of development projects ... In fact, without changing the development paradigm, these expensive conferences, goals and agreements will only result in failure. Development must not be reduced to 'growth', and 'construction'.

The portrayal of a counter position on the climate fund and international climate policy, i.e., demanding the fund, was fairly invisible in quantitative terms. Such invisibility indicates that this was a marginal viewpoint consistent with eco-socialist discourse "both in terms of space it took up in the media and the degree to which it was even mentioned in climate debate" (Anshelm & Hultman 2014, p. 85). This position was underpinned by questions raised about climate governance in Bangladesh (Sovacool 2017). However, this does not suggest that *Prothom Alo* or *The Daily Star* opposed the demands of government and civil society groups for the climate fund. Rather, they challenged the unilateral accusation levelled at the developed countries. In other words, they were critical of the view that only the developed countries were causing emissions and, by extension, climate change in Bangladesh. As suggested earlier, the newspapers were not necessarily against Bangladesh's climate change policy stance; they just wanted to remind everyone that Bangladesh should be careful about possible massive failure of its own environmental protection initiatives. In the main, editorial topics aspired to invoke a cosmopolitan vision (Beck 2015) that required rich developed nations to act for the common universal purpose of tackling climate change, compensating for their past emissions and making way for climate vulnerable countries, such as Bangladesh, to deescalate the process of climate change. Although it is fair to say that the coverage in the Bangladeshi publications did not include climate scepticism, there was evidence of solution scepticism, as described earlier in relation to critical domestication, in which questions were raised about development paradigms that were represented as "climate risks" for future catastrophes. This viewpoint was argued fairly thoroughly with the use of local and international evidence.

Conclusion: Economic rationality vs ecological vulnerability

The analysis of editorial commentaries pertinent to ongoing debates on climate change in both countries generated important insights, such as how various "presuppositions" (Develotte & Rechniewski 2001, p. 13) stemming from the absence or relative neglect of certain topics and excess attention to others were represented in the media politics of climate change (Beck 2006;

Cottle 2013, p. 1). For example, the low level of attention to global climate solidarity in Australia and to the development paradigm in Bangladesh was essentially underpinned by growing ethnocentrism or nationalism (Billig 1995; Olausson & Berglez 2014). The characteristics of this reduced emphasis differed in the two countries, but the Australian editorial commentaries seemed to have produced more persuasive arguments than those in the Bangladeshi publications. While the Australian commentaries tended to directly tell their constituents what to think about certain climate issues, Bangladeshi newspapers produced less persuasive but more informative and interrogative editorial positions.

Bangladeshi newspapers also appeared to have less coverage than those from Australia. This finding is consistent with some previous investigations (Painter 2017; Nassanga et al. 2017). The amount of coverage can be attributed to the fact that newspapers operate in the context of particular ideological conventions, such as liberal democratic or market value, and various taken-for-granted assumptions regarding the relationship between the media and political systems (Olausson & Berglez 2014). Perhaps Australia's carbon-intensive economy and an influential media system engendered more attention to the issue of climate change. Both coverage patterns together lead us to the questions of justice and solidarity that are crucial undercurrents in both journalism and global climate change (Roosvall 2017; Ward 2018). Opinion pieces from these two countries raised questions about how political thought, such as notions of justice or utilitarian thinking, intersects with the ethics and norms that should govern the responsible use of media space. Such responsible use includes informing or advocating on issues of public importance.

The editorials from the Australian publications defined climate change in economic rationalist terms, influenced by the dominant political actors (i.e., Liberal–National Conservative government) and directly shaped by considerations of economic costs and benefits alongside moral and scientific concerns. Economic rationality (Bulkeley 2000, 2001) has long played a key role in climate change politics, both nationally and internationally. In 2009, this economic rationality created uncertainty and antagonism in both the national and international contexts. However, in 2015, this economic rationality invoked a moment of "metamorphosis" (Beck 2015) that brought new opportunities for the emergence of "mission innovation" or "business-driven transformation" in Australia (i.e., CCS and RET). In other words, there was enthusiasm for solving climatic problems through technological innovations.

The coverage from Bangladesh also displayed a sort of "metamorphosis," or emancipatory side effect of global risk. Bangladeshi commentaries continued to claim compensation as the country was highly vulnerable to climate change. It is possible that this sustained political attention to the pursuit of climate justice in the news media caused their focus to shift from scrutinising existing local climate adaptation projects (Bhuiyan 2015). At the same time, the news media tended to see Bangladesh as a model or "aspiring example" of climate adaptation for the rest of the world. Such a vision for the country was not evident in the Australian commentaries, at least during the study periods. The analysis suggests that the two countries are located on different points in the spectrum of global risks pertinent to climate change. Questions pertaining to the producers of risks and those who are affected by them enabled critics to step out of the nation-state perspective and take a "cosmopolitan perspective" (Beck 2015). This perspective, which makes a significant contribution to understanding the "phenomenon of climate change and our changing experiences of weather" (Beck 2006; Hulme 2009), can provide valuable insights into the variations in the two countries' media coverage. Here, for instance, the concept of metamorphosis in Beck's cosmopolitan perspective is potentially useful. Based on the evidence from both countries, it seems plausible to argue that Australian commentaries were inclined to invoke the cosmopolitan perspective, partly because they envisioned a global climate solution through business-driven transformation and paid no attention to climate solidarity. The relevance of political cosmopolitanism (Held 2009; Robertson 2010), evident here in the case of Australia, but the absence of reference to global solidarity in relation to a truly global problem suggests that this perspective should be considered as an elitist project with a top-down trajectory.

A closer look at these two diverse countries, particularly the juxtaposition of high-emission Australia with low-emission Bangladesh, enhances understanding of the climate injustices inflicted on the latter. Arguably, a single-country, stand-alone assessment would have not generated this insight. However, following Scandrett (2007), it can be asserted that climate injustice is not so much something that is discovered from any specific research project, but something that is constructed through the contestations among various actors and processes, of which research is but a part (Beck 2010; Hulme 2009). In representing the respective positions, it is not important to determine which of the two discourses (economic rationality or climatic vulnerability) is more valid. Rather, it is crucial to identify their political implications. The validity of the discourses could be tested in different ways. It is also pertinent to ask what role a particular discourse plays in the journalistic practice that contributes to global climate change deliberations. From a normative perspective, editorial commentaries in Australia should demonstrate a greater level of solidarity with the issues around climate justice. Climate interpretations in this country's newspapers are underpinned by economic rationalism and the debates are dominated by questions about who produces climate risks (businesses, politicians, etc.). As a result, climate sceptic discourses sometimes find their way into the economic rationality debates. A greater focus on climate justice may help to subject the sceptic discourses to necessary rigorous editorial scrutiny. In the same spirit, it can be argued that the Bangladeshi editorials would benefit by achieving a more balanced focus between global climate justice issues and local adaptation efforts, as both are important for achieving a sustainable future for this vulnerable country.

References

Allan, S. 2002, *News culture*, Open University Press, Maidenhead.

Allan S., Adam, B. & Carter, C. 1999, 'Introduction: The media politics of environmental risk', in S. Allan, B. Adam, & C. Carter (eds) *Environmental risks and the media*, Routledge, London, pp. 1–26.

Anderson, A. 1997, *Media, culture and the environment*, UCL Press, London.

Anshelm, J. & Hultman, H. 2014, *Discourses of global climate change: Apocalyptic framing and political antagonisms*, Routledge, New York.

Bacon, W. 2011, *A sceptical climate: Media coverage of climate change in Australia 2011*, Australian Centre for Independent Journalism, Sydney.

Bacon, W. & Nash, C. 2010, 'Playing the media game: the relative (in)visibility of coal industry interests in media reporting of coal as a climate change issue in Australia', paper presented at the pre-ECREA Conference: Communicating climate change II, Institute of Journalism and Communication at the University of Hamburg, Hamburg, Germany.

Bauman, R. & Briggs, L. C. 1990, 'Poetics and performance as critical perspectives on language and social life', *Annual Review of Anthropology*, vol. 19, pp. 59–88.

Beck, U. 1992, *Risk society: Towards a new modernity*, translated by M. Ritter, Sage, London.

Beck, U. 2006, *Cosmopolitan vision*, Polity, Cambridge.

Beck, U. 2010, 'Climate for change or how to create a green modernity', *Theory Culture & Society*, vol. 27, no. 2–3, pp. 254–266.

Beck, U. 2015, 'Emancipatory catastrophism: What does it mean to climate change and risk society?', *Current Sociology*, vol. 63, no. 1, pp. 75–88.

Beck, U. 2016, *The metamorphosis of the world*, Polity, Cambridge.

Becker, H. S. 1967, 'Whose side are we on?', *Social Problems*, vol. 14, no. 3, pp. 239–247.

Bhuiyan, S. 2015, 'Adapting to climate change in Bangladesh: Good governance barriers', *South Asia Research*, vol. 35, no. 3, pp. 349–367.

Billett, S. 2010, 'Dividing climate change: global warming in the Indian mass media', *Climatic Change*, vol. 99, no. 1–2, pp. 1–16.

Billig, M. 1995, *Banal nationalism*, Sage, London.

Bulkeley, H. 2000, 'Discourse coalitions and the Australian climate change policy network', *Environment and Planning: Government and Policy*, vol. 18, no. 6, pp. 727–748.

Bulkeley, H. 2001, 'Governing climate change: the politics of risk society?', *Transactions of the Institute of British Geographers*, vol. 26, no. 4, pp. 430–447.

Carragee, M. K. 1993, 'A critical evaluation of debates examining the media hegemony thesis', *Western Journal of Communication*, vol. 57, no. 3, pp. 330–348.

Christodoulou, M. 2019, 'Gridlock: Radio script', Australian Broadcasting Corporation Radio National, Background Briefing, accessed February 12, 2019, available: https://www.abc.net.au/radionational/programs/backgroundbriefing/energy-policy-inaction-sparks-business-uncertainty/10766582.

Chubb, P. & Bacon, W. 2010, 'Australia: Fiery politics and extreme events,' in E. Eide, R. Kunelius & K. Kumpu (eds) *Global climate, local journalism: A transnational study of how media makes sense of climate summits*, Projekt Verlag, Freiberg, Germany, pp. 51-65.

Cottle, S. 2008, *Global crisis reporting*, McGraw-Hill Education, Maidenhead.

Cottle, S. 2010, 'Global crises and world news ecology', in S. Allen (ed.) *The Routledge companion to news and journalism*, Routledge, London, pp. 473–484.

Cottle, S. 2013, 'Environmental conflict in a global, media age: Beyond dualism', in L. Lester & B. Hutchins (eds) *Environmental conflict and the media*, Peter Lang, Oxford, pp. 19–33.

Das, J., 2015, 'Comparing journalisms: Newspaper coverage of river issues and climate change in Australia and Bangladesh', Doctoral thesis, University of Technology Sydney, Australia.

Dastagir, M. R. 2015, 'Modelling recent climate change induced extreme events in Bangladesh: a review', *Weather and Climate Extremes*, vol. 7, pp. 49–60.

Develotte, C. & Rechniewski, E. 2001, 'Discourse analysis of newspaper headlines: a methodological framework for research into national representations', *Web Journal of French Media Studies*, vol. 4, no. 1, pp. 1–13.

Eide, E. & Kunelius, R. 2010, 'Domesticating global moments', in E. Eide, R. Kunelius & K. Kumpu (eds) *Global climate, local journalism: A transnational study of how media makes sense of climate summits*, Projekt Verlag, Freiberg, Germany, pp. 11–40.

Eide, E., Kunelius, R. & Kumpu, K. 2010, *Global climate, local journalism: A transnational study of how media makes sense of climate summits*, Projekt Verlag, Freiberg, Germany.

Fairclough, N. 1992, *Discourse and social change*, Polity, Oxford.

Fairclough, N. 2003, *Analysing discourse: Textual Analysis for social research*, Routledge, London.

Farbotko, C. 2005, 'Tuvalu and climate change: constructions of environmental displacement in the Sydney Morning Herald', *Geografiska Annaler: Series B, Human Geography*, vol. 87, no. 4, pp. 279–293.

Gal, S. 2006, 'Linguistic anthropology', in K. Brown (ed.) *Encyclopedia of language and linguistics*, Elsevier, Oxford, pp. 171–185.

Giddens, A. 2011, *The politics of climate change*, Polity Press, Cambridge.

Goodman, J. 2016, 'The "climate dialectic" in energy policy: Germany and India compared', *Energy Policy*, vol. 99, pp. 184–193.

Grantham Research Institute on Climate Change and Environment 2018, 'Bangladesh country profile', The London School of Economics & Politics, accessed February 12, 2019, available: http://www.lse.ac.uk/GranthamInstitute/country-profiles/bangladesh/.

Guha, R. & Martinez-Alier, J. 1997, *Varieties of environmentalism: Essays north and south*, Earthscan, London.

Hanlon, J., Roy, M. & Hulme, D., 2016, *Bangladesh confronts climate change: Keeping our heads above water*, Anthem Press, London.

Harvey, D. 1993, 'From space to place and back again: Reflections on the condition of postmodernity', in J. Bird, B. Curtis, T. Putnam & L. Tickner (eds) *Mapping the futures: Local cultures, global change*, Routledge, London, pp. 3–39.

Held, D. 2009, 'Restructuring global governance: Cosmopolitanism, democracy and the global order', *Millennium*, vol. 37, no. 3, pp. 535–547.

Hindman, E. B. 2003, 'The princess and the paparazzi: Blame, responsibility, and the media's role in Diana's death', *Journalism & Mass Communication Quarterly*, vol. 80, no. 3, pp. 666–688.

Hodges, A. 2015, 'Intertextuality in discourse', in D. Tannen, H. E. Hamilton & D. Schiffrin (eds) *The handbook of discourse analysis*, John Wiley & Sons, Chichester.

Huckin, T. N. 1997, 'Critical discourse analysis', in T. Miller (ed.) *Functional approaches to written text*, US Department of State, Washington, DC, pp. 78–92.

Hulme, M. 2009, *Why we disagree about climate change: Understanding controversy, inaction and opportunity*, Cambridge University Press, Cambridge.

Hulme, M. 2010, 'Cosmopolitan climates: Hybridity, foresight and meaning', *Theory, Culture & Society*, vol. 27, no. 2, pp. 267–276.

Huq, S. 2015, 'The inside story of the Paris agreement', *The Daily Star*, December 15, accessed February 12, 2018, available: https://www.thedailystar.net/op-ed/the-insi de-story-the-paris-agreement-187159.

Huq, S. 2019, 'Changing the climate narrative', *The Daily Star*, January 2, accessed February 12, 2018, available: https://www.thedailystar.net/opinion/politics-clima te-change/news/changing-the-climate-narrative-1681573.

Inman, M. 2009, 'Where warming hits hard', *Nature Reports Climate Change*, January 15, pp. 18–21, accessed December 20, 2018, available: https://www.nature.com/arti cles/climate.2009.3.

Intergovernmental Panel on Climate Change (IPCC) 2007, *General guideline on the use of scenario data for climate impact and adaptation assessment*, Task Group on Data and Scenario Support for Impact and Climate Assessment (TGICA), IPCC, Geneva, Switzerland.

Intergovernmental Panel on Climate Change (IPCC) 2014, 'Climate change 2014: Synthesis report', in R. K. Pachauri & L. A. Meyer (eds) *Contribution of Working Groups I, II and III to the Fifth Assessment Report of the Intergovernmental Panel on Climate Change*, IPCC, Geneva, Switzerland.

Islam, M. S. & Islam, M. N. 2016, 'Environmentalism of the poor: the Tipaimukh dam, ecological disasters and environmental resistance beyond borders', *Bandung: Journal of the Global South*, vol. 3, no. 1, pp. 1–16.

Johnson, A. 2017, *Bangladesh's climate change challenge*, The National Bureau for Asian Research, accessed May 18, 2018, available: https://www.nbr.org/publication/ bangladeshs-climate-change-challenge/.

Kovach, B. 2006, 'Toward a new journalism with verification', *Nieman Reports*, vol. 60, no. 4, p. 39.

Kunelius, R. 2012, 'Varieties of realism: Durban editorials and the discursive landscape of global climate politics', in E. Eide & R. Kunelius (eds) *Media meets climate: The global challenge for journalism*, NORDICOM, Göteborg, Sweden, pp. 31–48.

Kunelius, R. & Eide, E. 2017, 'The problem: Climate change, politics and the media', in R. Kunelius, E. Eide, M. Tegelberg& D. Yagodin (eds) *Media and global climate knowledge*, Palgrave Macmillan, New York, pp. 1–32. Lemly, A. D. 2017, 'Environmental hazard assessment of coal ash disposal at the proposed Rampal power plant', *Human and Ecological Risk Assessment: An International Journal*, vol. 24, no. 3, pp. 627–641.

Lubcke, T. 2013, 'Will the opposition's direct action plan work?', *The Conversation*, accessed August 10, 2018, available: https://theconversation.com/will-the-opposi tions-direct-action-plan-work-12309.

McGregor, S. 2003, 'Critical discourse analysis – A primer', *Kappa Omicron Nu Forum*, vol. 15, no. 1, accessed November 11, 2018, available: http://www.kon.org/a rchives/forum/15-1/mcgregorcda.html.

McNair, B. 2014, 'Hard news: the carbon tax shows up cracks in media reporting', *The Conversation*, accessed December 14, 2018, available: https://theconversation. com/hard-news-the-carbon-tax-shows-up-cracks-in-media-reporting-29206.

Meyer, K. E. 2001, '150th anniversary: 1851–2001; Dept. of conscience: The edi torial "we"', *New York Times*, November 14, accessed January 10, 2019, avail able: https://www.nytimes.com/2001/11/14/news/150th-anniversary-1851-2001-dep t-of-conscience-the-editorial-we.html.

Nash, C. 2016, *What is journalism? The art and politics of a rupture*, Palgrave Macmillan, London.

Nassanga, G., Eide, E., Hahn, O., Rhaman, M., Sarwono, B. 2017, 'Climate change and development journalism in the global south', in E. Eide, R. Kunelius & K. Kumpu (eds) *Global climate, local journalism: A transnational study of how media makes sense of climate summits*, Projekt Verlag, Freiberg, Germany, pp. 213–233.

Neverla, I. 2008, 'The IPCC reports 1990–2007 in the media: A case study on the dialectics between journalism and natural science', paper presented at Global Communication and Social Change, International Communication Association (ACA) conference, Montreal, Canada, May 22–26.

Nord, D. P. 2001, *Communities of journalism*, University of Illinois Press, Urbana, IL.

Olausson, U. 2009, 'Global warming—global responsibility? Media frames of collective action and scientific certainty,' *Public Understanding of Science*, vol. 18, no. 4, pp. 421–436.

Olausson, U. 2014, 'The diversified nature of "domesticated" news discourse: The case of climate change in national news media', *Journalism Studies*, vol. 15, no. 6, pp. 711–725.

Olausson, U. & Berglez, P. 2014, 'Media and climate change: Four long-standing research challenges revisited,' *Environmental Communication*, vol. 8, no. 2, pp. 249–265.

Painter, J. 2013, *Climate change in the media*, I. B. Taurus, London.

Painter, J. 2017, 'Disaster, risk or opportunity? A ten-country comparison of themes in coverage of the IPCC AR5', in E. Eide, R. Kunelius & K. Kumpu (eds) *Global climate, local journalism: A transnational study of how media makes sense of climate summits*, Projekt Verlag, Freiberg, Germany, pp. 109–128.

Pan, Z. & Kosicki, G. 1993, 'Framing analysis: An approach to news discourse', *Political Communication*, vol. 10, no. 1, pp. 55–75.

Patterson, T. E. 2013, *Informing the news: The need for knowledge-based journalism*, Vintage Books, New York.

Rahman, M. A. 2018, 'Governance matters: climate change, corruption, and livelihoods in Bangladesh', *Climatic Change*, vol. 147, no. 1–2, pp. 313–326.

Roberts, J. T. & Parks, B. 2007, *A climate of injustice: Global inequality, north-south politics, and climate policy*, MIT Press, Boston, MA.

Robertson, A. 2010, *Mediated cosmopolitanism: The world of television news*, Polity, Cambridge.

Roosvall, A. 2017, 'Journalism, climate change, justice and solidarity: Editorializing the IPCC AR5', in E. Eide, R. Kunelius & K. Kumpu (eds) *Global climate, local journalism: A transnational study of how media makes sense of climate summits*, Projekt Verlag, Freiberg, Germany, pp. 129–150.

Rupar, V. 2007, 'Newspapers' production of common sense: The "greenie madness" or why should we read editorials?', *Journalism*, vol. 8, no. 5, pp. 591–610.

Said, E. 1978, *Orientalism*, Routledge & Kegan Paul, London.

Scandrett, E. 2007, 'Environmental justice in Scotland: policy, pedagogy and praxis', *Environmental Research Letters*, vol. 2, no. 4, 045002.

Schmidt, A., Ivanova, A., & Schäfer, M. 2013, 'Media attention for climate change around the world: Data from 27 countries', *Global Environmental Change*, vol. 23, pp. 1233–1248.

Shanahan, M. 2009, 'Time to adapt? Media coverage of climate change in non-industrialised countries', in T. Boyce & J. Lewis (eds) *Climate change and the media*, Peter Lang, New York, pp. 145–157.

Smith, B. & West, P. 1996, 'Drought, discourse, and Durkheim: A research note,' *Australia & New Zealand Journal of Statistics*, vol. 32, no. 1, pp. 93–102.

Sovacool, B. K. 2017, 'Bamboo eating bandits: Conflict, inequality and vulnerability in the political ecology of climate change adaptation in Bangladesh', *World Development*, vol. 102, pp. 183–194.

van Dijk, T. A. 1988, *News as discourse*, Lawrence Erlbaum, Hillsdale, NJ.

van Dijk, T. A. 1991, *Racism and the press*, Routledge, London.

van Dijk, T. A. 1993, 'Principles of critical discourse analysis', *Discourse & Society*, vol. 4, no. 2, pp. 249–283.

van Dijk, T. A. & Kintsch, W. 1983, *Strategies of discourse comprehension*, Academic Press, New York.

Waisbord, S. 2010, 'Rethinking "development" journalism', in S. Allan (ed.) *The Routledge companion to news and journalism*, Routledge, London, pp. 148–158.

Ward, S. J. 2018, *Ethical journalism in a populist age: The democratically engaged journalist*, Rowman & Littlefield, Lanham, MD.

Wodak, R. & Fairclough, N. 1997, 'Critical discourse analysis', in T. A. van Dijk (ed.) *Discourses as social interaction*, Sage, London, pp. 258–284.

Ytterstad, A. 2014, 'Framing global warming: Is that really the question? A realist, Gramscian critique of the framing paradigm in media and communication research', *Environmental Communication*, vol. 9, no. 1, pp. 1–19.

Zelizer, B. 2004, *Taking journalism seriously: News and the academy*, Sage, London.

5 Sources of Australian climate change news

In the previous chapter, the elements that shaped editorial commentaries in the two countries were mapped. In journalism, editorial commentaries are written in what is known as a reflective style of writing. The commentaries examined in the previous chapter articulated the reality of climate change "in the ideological framework" (Broersma 2010, p. 29) of different social and political groups. In contrast, the news style deals with facts and information and uses its own particular style. This chapter focuses on the latter. It has been argued that journalists do not simply report events, but use sources to frame and interpret news to influence public perceptions (van Dijk 1991; Tuchman 1978). This chapter is concerned with how climate change was reported in *The Sydney Morning Herald* and *The Australian* in 2009 and 2015, with particular focus on how various sources were called upon. The analysis considers the journalistic strategies employed in the selection and representation of sources. The flows of information between diverse interest groups, such as politicians, scientists, conservationists-cum-activists, businesses and farmers, were examined to establish how news discourses that claim to "represent social reality" (Peters & Broersma 2017, p. 189) were produced through the use of these groups. The chapter also seeks to ascertain how far the news media were able to uphold their critical gaze (Berkowitz 2009) in the coverage. This exploration provides insights into the "nature of news as a form of knowledge and journalism as a knowledge acquiring discipline using the logical empirical principles of social sciences" (Park 1940, quoted in Pan & Kosicki 1993, p. 61). In the process of knowledge making, journalists or news makers engage in hypothesis testing through the use of direct observation or direct quotes from sources, thereby engaging in inductive and deductive reasoning on the issue of climate change. An analysis of these sources helps to reveal the extent to which journalists are inclined to check or verify statements from their principal sources. As discussed in Chapter 3, principal sources are individual news sources (e.g., politicians, bureaucrats, experts or citizens) who provide journalists with important information or statements in support of or against the main theme of an article. Checking refers to a critical examination of contestable statements made by sources, or the inclusion of "another voice with another opinion"

(Diekerhof & Bakker 2012, p. 248) in the news texts. To determine whether or not an article has been checked, it is carefully scrutinised for signs of critical examination, such as cross-checking of facts attributed to principal sources with other sources, or questioning of the logical plausibility of the sources' claims. The cross-checking of excerpts can be considered as one of the closest possible indicators of verification in the selected news text.

The news text is "a system of organized signifying elements" (Pan & Kosicki 1993, p. 55) in which sources are used to argue certain points over others to test a particular hypothesis related to climate change. In this hypothesis-testing process, the focus is on a "topic at a time ... where events are cited, sources or actors are quoted and propositions are pronounced" (Pan & Kosicki 1993, p. 60) in order to establish a frame presented or implied, based on evidence from the journalists' quotation of sources, and background information (Broersma 2010). In this analysis, frame is perceived as a tool for argument and counter argument (Jönsson 2011).

Sources in climate change news

The news coverage of climate change leading up to the two summits—the 2009 Conference of Parties in Copenhagen (COP15) and the 2015 meeting in Paris (COP21)—was heavily dominated by political sources ($n = 888$), followed by experts ($n = 331$), bureaucrats ($n = 215$), businesses ($n = 232$) and activists ($n = 128$). These figures are useful to begin the discussion, but not adequate to provide a full picture of the "determinants and dynamics involved" (Cottle 2009, p. 72) in the reporting of climate change. This section presents an analysis of news articles pertaining to climate change as well as the two summits (COP15 and COP21) collected during two periods of six months in 2009 and 2015. Particular attention is paid to the use and validation or cross-checking of different statements made by the principal sources. The findings demonstrate that there was a slight increase of representation of business sources in 2015 compared to 2009, both in total number and in the number of principal sources. The proportionate share of other sources, including experts and activists, increased during the period, but political and bureaucratic sources decreased. This shift in the use of sources, particularly the decrease in political sources, arguably indicates fewer uncertainties and a more action-oriented approach to climate policy in 2015 compared to the scenario in 2009. With this brief background, the following sections focus on the details pertinent to politicians and bureaucrats, two of the most powerful news sources who have high accessibility in the process of news production (Gans 1979; Cottle 2000; Strömbäck & Nord 2006), followed by business sources. The subsequent sections present a similar discussion of experts, who are an integral part of news work and provide journalists with crucial information for public benefit, followed by activist sources.

Political sources

The study found that there had been intense political debates before the climate conferences and there was an increased presence of political sources in the news coverage concerning both national and international issues. This section explores how these political sources were quoted "to provide meaning to an unfolding strip of events" (Gamson & Modigliani 1987, p. 143) and the extent to which quotations from these sources were checked to provide "interpretation" of the responsibility for the issues raised (Sreberny & Paterson 2004, p. 8). It also discusses the level of empirical support in the hypothesis testing feature of these pieces of journalism.

The international summits in Copenhagen and Paris were particularly significant for Australia as the national political debates surrounding the proposed Emissions Trading Scheme (ETS) in 2009 and Renewable Energy Target (RET) in 2015 gained traction in both periods. A climate summit has the capacity to promote "awareness of risk" by creating a global public; at the same time, it can present a unique challenge to global regulations (Beck 2009; Eide et al. 2010) on environmental issues. For global regulations to be effective, relevant policies have to be negotiated by the leaders of different countries. The focus in both 2009 and 2015 was on how politicians from both national and international contexts were portrayed or checked in the pages of the two newspapers. Climate summits and national climate policies were highly contested political matters. News articles made extensive reference to political leaders or political apparatus from the US, China and India in both periods, but the exercise of checking of source excerpts differed in the two study periods (see Appendix 2, Table A2.6). Here, the news headlines were considered as arguments and excerpts from the sources (Pan & Kosicki 1993) were considered as constituting elements of the arguments and counter-arguments that were used to frame the reporting of climate change. The selection and positioning of sources by editors and journalists also contribute to the frame building process (see Chapter 3).

The coverage in Australia in 2009 invoked conflict frame as a result of disagreement between the political leaders. In 2015, however, the frame took the form of an appeal to action that showed the unity of purpose (Pham & Nash 2017) among different political groups, particularly between politicians and the business community. The analysis of the articles, with particular focus on the positioning of source excerpts, shows how the arguments made in headlines were pursued. In 2009, on one side of the debate, articles (e.g., "Late bid to save climate talks—Hopes fade in Copenhagen, rise on the reef," *The Australian,* December 19, 2009) clearly identified the crux of the issue as the (Copenhagen Accord) which "fell far short of the decisive action" due to disagreement between the rich nations and China and other developing countries." This interpretation resonated in the speech of US President Barack Obama, the principal source of the article, who stated: "Mitigation, transparency, financing: it is a clear formula ... At this point, the question is whether we move forward together or split apart, whether we prefer posturing to action" (*The Australian,* December 19, 2009). The excerpt is significant, particularly in its use of the words "transparency" and "action," which were considered to be crucial to the signing of a climate deal. In accordance with this deal, the US was demanding that China should agree to "international verification of its pollution fighting efforts." Without such an arrangement, the US rejected the possibility of any deal at the Copenhagen Summit. China, however, rejected the notion of international verification outright on the grounds of its sovereignty. The article used an anonymous bureaucratic source to detail the meetings between the US and China and other developing nations. This source revealed that Premier Wen Jiabao was reluctant to engage in negotiations with US President Barack Obama. The reluctance was made clear when China was represented at this high-level meeting by a "third ranking official" whose presence the US termed "unhelpful." From this evidence—reference to the Chinese leader's rejection of the US conditions, the labelling of Chinese participation as "third ranking," and failure to include any significant Chinese explanation of their reluctance to sign a binding climate treaty, it may be deduced that the article played up the conflict theme by representing the whole issue as a contestation between the developed nations led by Barack Obama and the developing nations led by Wen Jiabao.

Despite disagreement among leaders, the coverage in *The Sydney Morning Herald* (e.g., "Blessed or blamed? A little of both," December 19, 2009) portrayed the summit as a symbol of "unstoppable momentum" and focused on addressing global climate change. Yet, the articles from this newspaper also held China responsible for the failure of the negotiations, albeit in conjunction with the US. In this context, the article cited the Swedish Environment Minister Andreas Carlgen who directly blamed the US and China for failing to initiate an "attempt to halt dangerous climate change." As suggested earlier, a significant aspect of the articles was the inclusion of diverse views, which not only reflected the viewpoints of the powerful world political leaders, but also signalled the apprehension of third world leaders regarding climate change negotiations. For example, Bangladesh's Prime Minister Sheikh Hasina spoke of the consequences of a rise in sea level for her country. Ethiopia's Prime Minister Meles Zenawi expressed his "frustration" at the increasing number of deaths from famine across Africa. These potential catastrophes were linked to the wealth and wellbeing enjoyed in developed countries through carbon-intensive development which was "fundamentally unjust." The portrayal of their concerns by various political leaders, including Danish and third world politicians, resonated in a comment by Danish Climate Change Minister Connie Hedegaard: "We are all accountable. Not only for what we do but also for what we fail to do." The selection and inclusion of this direct quote from Danish Minister regarding the prospects of the conference, particularly the phrase "all accountable," could be understood as an attempt to hold politicians from different parts of the world responsible for the Summit's failure. It can also be argued that both sides of the debate reflected in articles in *The Sydney Morning Herald* and *The Australian* played up the conflict between developed and developing nations by emphasising the "concerns" of the third world leaders and the "frustration" of the conference hosts (Danish politicians). However, this conflict (Nisbet 2010) was characterised differently from that of the article in *The Australian* which directly positioned the willingness of the US and the reluctance of China to negotiate a deal. The article in *The Sydney Morning Herald* argued that both China and the US were responsible for engendering the deadlock by emphasising the concerns of the developing countries (G77), and the apprehension expressed by the Danish leader.

Unlike in 2009, during the 2015 Paris Summit, both national and international news articles attributed responsibility for climate change to the Australian government (Iyengar 1996). Headlines and the positioning of sources and associated excerpts focused on the Australian government, which was assumed to have the ability to address the issues surrounding climate change. In 2015, the articles invoked an action frame, highlighting the common interest emerging among various political groups and business bodies, although with some reservations. On one side, particularly in the case of the Paris conference, a sense of unity and commitment among world leaders and business communities (see also Painter 2016) was demonstrated with the formation of a "Breakthrough Energy Coalition" (Arup & Kenny 2015), that opened up a new "trillion-dollar opportunity" (Walsh 2016) rivalling the industrial revolution and ushering in new partnerships between politicians and businesses. On the other side, Australia's commitment to the Paris Agreement put the spotlight on the national debate on climate policies (e.g., emissions reduction target) because of the sluggish progress that had resulted from political contestation around the costs and benefits of the proposed measures (Holmes 2015).

Broadly, the articles that were dominated by political sources argued that reaching consensus on a plausible climate deal would be a good thing, creating avenues for new business opportunities; as one article stated: "World unites on climate but critics wanted more" (*The Australian* December 14, 2015). This article used six sources, four political and two business sources, and described Australia's positive role in the historic international climate agreement in Paris that required all countries to act. The principal source, Foreign Minister Julie Bishop, was cross-checked with Greens leader Richard Di-Natale, who stated: "the Paris agreement would be a success only if countries including Australia seriously ratcheted up their targets between now and when the new agreement takes effect from 2020." Political leaders from across the world, however, perceived the conference as a success. The Indian Prime Minister Norendra Modi, who outlined a renewable energy programme with France, "described the deal as a victory of 'climate justice' and said there were no winners or losers in the outcome. 'Climate Justice has won and we are all working towards a greener future' Modi tweeted." The expression "we … all" could be indicative of the call for solidarity and global responsibility in tackling climate change, which remained a crucial aspect of international climate politics (Okereke & Coventry 2016; Roosvall 2017). The main theme of the article was the unity between politicians and the business community; for example, business representatives, such as the Australian Industry Group chief executive Innes Willox, agreed with Australia's commitment to the Paris Agreement but emphasised the need for "stable, flexible and cost-effective policies." Similarly, economic concern was raised via the words of Foreign Minister Julie Bishop, who stated: "We need to have the right balance between the economy and environmental outcomes and believe this (Paris) agreement provides the framework for countries to do precisely that."

Nevertheless, opposition political forces such as the Labor Party were used to demonstrate how the "Agreement puts heat on PM's domestic policies" (*The Sydney Morning Herald*, December 14, 2015). The article also used Foreign Minister Julie Bishop, but she was cross-checked by the opposition Labor Party's environment spokesperson, Mark Butler, who thought the government still had a lot to do: "Mr Turnbull's policies, including pollution reduction targets and an intention to abolish several climate-related agencies, were 'massively out of step with the rest of the world and completely inconsistent with the agreement that was struck overnight.'" However, it also quoted Mineral Council Australia's chief executive Brendan Pearson:

> Paris agreement would support low-emissions coal technologies. The deal would aid the development of carbon capture and storage and support the 'rapid growth of new generation nuclear power in East and South Asia, providing strong demand for Australian uranium exports.'

The positioning of these excerpts from the opposition and business sources can be interpreted as the journalistic inclination to highlight technological solutions to climate change despite political differences between the government and opposition. Although political sources dominated this debate, the journalistic style sought to persuade readers of the feasibility of technological or business solutions to climate change, albeit at the expense of global solidarity or global justice, which remained relatively invisible.

Billion dollar question

Although government ministers, such as Julie Bishop, were used to highlight the Paris framework's capacity to provide various countries' pledges, some country pledges received negative press attention. A case in point: "$138bn a year aid fund is 'a floor, not a ceiling'" (*The Australian,* December 14, 2015). It stated: "A $US100 billion ($138bn) a year fund to help developing countries cope with climate change will be boosted from 2025, with all parties invited to contribute on a voluntary basis." The article was cross-checked with the Minister for International Development and the Pacific, Steve Ciobo, who confirmed the current commitment of $200m and rejected the opposition Labor Party leader

Bill Shorten's suggestion that money could be stripped from child mortality or literacy programs, saying projects would be funded in consultation with the requests of the recipient countries. He stressed the $200m annual commitment was 'a floor, not a ceiling.'

This expression reflects a strongly ethnocentric position in which Australia is cast in a positive light because of its support for the welfare of development partners. The stronger the floor, the safer the funded country would be. Although a number of political sources (the Australian Greens and the Labor Party) were used in relation to climate aid for developing countries, their positions were not highlighted in the articles dominated by political sources. In these articles, they were used mainly as secondary sources. As critics have argued, due to "deadline pressure and the significance of obtaining information rich in news values" (Bennett 2003, p. 125), journalists are heavily dependent on bureaucratic sources, another group of significant participants in the political process (Ericson et al. 1989). Bureaucratic sources frequently appear as news sources in articles concerning key international events (Gitlin 2003; Lawrence 2000; Shehata 2007; Wolfsfeld & Sheafer 2006), and this was the case in the present analysis, particularly in relation to climate aid.

The topic of climate aid gained significant traction in the 2009 coverage, where news makers positioned both the national and international bureaucrats "to stage" (Ericson et al. 1991, p. 4) the Copenhagen and Paris summits globally and emissions reduction policies locally. In quantitative terms, the number of bureaucratic sources—both principal and secondary—were similar in the two newspapers. However, the verification of bureaucrats as principal sources, which characterises journalistic knowledge (Kovach & Rosenstiel 2007; Zelizer & Allan 2010), differed between the two periods. While Copenhagen was marred by leaked documents (such as the Danish draft, discussed below), Paris was a regular international meeting. News articles concerning the Paris summit mostly used international bureaucrats to highlight the debate surrounding Australian climate policy, which was directly related to Australia's commitment to emissions reduction.

Bureaucratic sources

During the Copenhagen summit, some articles ("Late bid to save climate talks—Hopes fade in Copenhagen," December 19, 2009, *The Australian*; "Blessed or blamed? A little of both," December 19, 2009, *The Sydney Morning Herald*) highlighted the reactions of bureaucratic sources from developing nations to the leak of the controversial Danish draft on the eve of the conference. The leaked draft document revealed the goal of limiting global warming by 2°C by developing and developed countries, as well as a $10-billion-a-year climate aid package for the former, but without any specific commitment regarding aid. While the stories described the positions of the developing nations in response to the leaked document, there were differences in how the sources were quoted. The principal source of the article titled "Climate deal backers 'like Nazi appeasers'" (*The Australian,* December 10, 2009) was the Sudanese negotiator, Stanislaus Di-Aping, who described the draft document as a "new form of rich country imperialism designed to divide poor nations and maintain the dominance of the developed world." Nonetheless, in the same report, other bureaucratic sources, such as the Tuvalu representative, were also cited, demanding a potential emissions reduction deal from the positions of their respective countries. China's chief climate negotiator, Su Wei, for example, was not particularly concerned about the commitment to climate aid to developing countries. Rather, she held the rich countries directly responsible for making an insignificant commitment to emissions reduction.

The notion of a rich–poor divide again appeared in an article titled "Turning up the heat in Copenhagen" (*The Sydney Morning Herald,* December 10, 2009), in which Di-Aping described the Western offer of aid as "not enough to buy us coffins." His comment was directly refuted by the European Union official, Artur Runge-Metzger, who commented that "Di-Aping was unaware of all European aid effort in Africa because he lived in New York." This article also emphasised that differences of opinion did not exist solely in the developing nations. A case in point here was the inclusion of comments made by the Australian and Danish Climate Change ministers in relation to the progress of the conference, comments that reflected the different standpoints of these two countries. The Danish Environment Minister, Connie Hedegaard, tried to mobilise her third world counterparts by asserting that "This is the time to deliver! This is the place to commit! Let's get it done!" Her Australian counterpart was quoted as saying that "the negotiation was 'difficult' and language around it 'unhelpful.'" Thus, the first world politicians viewed the climate negotiations as both "deliverable" and "difficult," while the third world representatives pressed their demand for a binding climate treaty with "fierce urgency."

The discussion surrounding climate aid also surfaced during the Paris summit, where it was referred to using the bureaucratic label "climate finance." The coverage, however, did not engage with bureaucrats from developing nations or climate vulnerable states as it had done in 2009. The issue of climate assistance in the context of the Australian government's foreign aid cut and subsequent collaboration with the private sector as aid "partners" (McDonald 2015) was acknowledged from two different perspectives. First, Australia contributes adequately to the International Climate Fund; second, future funding (e.g., $100bn US fund) is contingent on the needs of developing countries as well as their assurance to rich countries that they (developing countries) would relinquish the right to claim any damages or compensation for global warming. Referring to Chief US negotiator, Todd Stern, the article stated: "There is one thing we don't accept and won't accept in this agreement and that is the notion that there should be liability and compensation for loss and damage" ("Sacrifice rights for deal on carbon,"

The Australian, December 9, 2015). This position was also supported by comments from Foreign Minister Julie Bishop. However, Climate Institute deputy director Erwin Jackson expressed concerns about inadequate financial contributions from Australia in *The Sydney Morning Herald* article titled "Agencies slam aid repackaging" (December 2, 2015). It used former Prime Minister Malcom Turnbull but highlighted Marc Purcell, chief of the Australian Council for International Development, who described the Australian government initiative as "deeply disappointing," adding: "Let's be clear; this is not new money. This is spending that has already been announced. While we welcome the government acknowledging climate change challenges faced by our Asia Pacific neighbours, this announcement lacks any ambition." Here, the selection of sources and their viewpoints on climate aid showed how some sources expressed solidarity with the imperatives of a climate fund for developing partners, but others clearly lacked the solidarity and sought to pre-empt consequences for rich countries in the near future. A comparison of the two periods demonstrates that, unlike Copenhagen, the agents or sources from climate vulnerable states were relatively invisible during the Paris summit. As a result, the articles in Australia offered a fragmented view of the aspired-to global responsibility (of much aspired global) that is strongly associated with the idea of a climate fund for vulnerable countries. In the case of climate aid, the two Australian newspapers held opposing positions, but in the official-source-dominated articles, a broad consensus was identified in relation to the Australian government policy of achieving the emissions reduction target to which it had committed in the Intended Nationally Determined Contribution (INDC).

The articles dominated by official sources broadly attributed responsibilities to the government, as both sides of the debate argued over the process of achieving the emissions reduction target. Within this argumentation process, both sides demonstrated a broad consensus over climate policy, albeit with some slight differences. The moment of broad consensus emerged with the release of the report by the Climate Change Authority (CCA). One of the articles outlined its concern: "'We have lost sight of the goal': calls for debate to find purpose" (*The Sydney Morning Herald*, December 1, 2015). In another article, *The Australian* argued: "Climate policy 'needs a fresh look'" (December 1, 2015). Both articles focused on the acting Chairman of the CCA, Stuart Allinson, and cross-checked with business sources in *The Australian* and political sources in The *Sydney Morning Herald*. The articles dominated by political sources were heavily focused on the contestation between the two sides of federal politics on the emissions reduction target as well as the emissions reduction scheme. Referring to the acting Chairman of the CCA, the article in the *The Australian* (December 1, 2015) described climate policy as "'highly polarised' and it appeared at times 'we have lost sight of the key goal' to cut emissions." It referred to a draft report by the CCA, which indicated that "meeting the government's target for 2030 of 26 to 28 per cent below 2005 levels 'is likely to remain a substantial task.'" The article also put forward the position from the business side by quoting Trevor St

Baker, founder of the market-listed energy company ERM Power and part-owner of the Vale Point Coal Power Station in NSW: "a target of 45 per cent cuts by 2030 compared with 2005 levels would cause 'mass job losses and business losses.'" Clearly, by using only a business source for checking, this article narrowed the perspective on climate policy and emissions reduction in support of an argument for a fresh approach to protect jobs and the economy.

The protection of the coal industry also gained traction through reference to International Energy Agency (IEA) policy, which suggested that coal-use should be commercially viable and, at the same time, climate policy compliant. *The Australian* highlighted this in a one-source exclusive: "Five-year deadline for clean coal: IEA" (*The Australian*, October 6, 2015). It referred to the IEA executive director, Dr Fatih Birol, who: "played down predictions by environmentalists of the end of the coal industry. 'To declare that coal is dead may be in my view a rather premature way of thinking,' he said." In relation to the increasing demand from India, the article stated that the country had 350 million people without access to electricity, which would underpin coal demand. It also focused on carbon capture and storage (CCS) technology, which "remained a most critical technology." Dr Birol was also used by *The Sydney Morning Herald* as a source in the background to the Paris agreement ("IEA takes axe to coal forecasts," December 19, 2015). The article stated: "Dr Birol lamented the 'terribly slow' process in carbon capture and storage, which he said was bad news for the environment, but is much worse news for the coal industry." In the process of validation, the article used a business source, Minerals Council executive director Brendan Pearson, who challenged the IEA prediction, stating that the "new forecast does not translate to a reduction in Chinese demand for high quality coal imports, including from Australia." Greenpeace Australia Pacific climate campaigner, Shani Tager, however, viewed the IEA position positively, and described this report as "another nail in coal's coffin." The positioning of Shani Tager's comment without any other evidence served the purpose of the article, which was to strengthen the position on action for emissions reduction. It also provided its readers with the other side of the argument by using the Mineral Council executive.

Both newspapers used political and bureaucratic sources as significant information providers and players in the Copenhagen climate change summit, but this was less prevalent during the Paris summit (see Appendix 2, Table A2.4). What is important here is not the use of principal and other sources in the process of cross-checking but, rather, the selection and positioning of quotes from certain sources over others. This allowed the newspapers to present a reliable version of facts about the issues in the climate change conferences, albeit one that differed between the two meetings (Swain & Robertson 1995). While the articles in *The Australian* depicted the international issue as a polarisation between rich developed nations and poor developing countries, *The Sydney Morning Herald* also saw it as a matter of contestation between the first and third worlds, a process in which various politicians were equally accountable for the impasse. Here, it is evident from the sources used for cross-checking that journalists substantiated their arguments on climate policy internationally and nationally and set up the parameters of a "cognitive window" by excluding certain sources and including others in the argumentation process.

A comparison of the two periods, particularly in relation to the COPs, showed that political and official sources were dominant, particularly in relation to the topic of climate aid. In the 2009 coverage, other voices, such as perspectives from the developing countries, were present but were negatively portrayed. In 2015, however, these sources from the developing world were invisible. The absence of other sources or other views further normalised the ethnocentric, narrowly national-interest-driven coverage in Australia. This raises questions about journalistic moral responsibility, which should take cognisance of human rights across the spectrum (see Ward 2018). During the Paris summit, this perspective was manifested in two ways: by failing to provide space for sources from developing countries; and by favouring business sources in cross-checking on matters concerning local coal policy.

Business sources

The presence of business sources in COP21 demonstrated a significant shift not only from the Copenhagen coverage but also from the overall strategies of covering the UN Framework Convention on Climate Change (UNFCC) in the 1990s. Back then, business sources were defensive and obstructive and raised challenges to climate science (Walsh 2016). The new approach involved a "business driven transformation" in coverage of the Paris meeting ("Climate change, the trillion-dollar industry to rival the industrial revolution," *The Australian*, 12 December, 12, 2015). In 2015, business sources were predominantly framed to demonstrate the policy impact on industry and the consequences for consumers or the Australian public (Renewable Energy Target and increase in power prices) as well as the ongoing effects of climate change (or warming weather on the wine and farming industry). In 2009, however, the newspaper debates adopted a negative stance towards the Australian government's policies and their effects on industries and ordinary people.

They also scrutinised business activities related to the implementation of various climate policies. A case in point here is the coverage of the global carbon scheme, which offered insights into the newspapers' articulation of "business interests" in their respective content (Ericson et al. 1989, p. 260). An article titled "Australian firm linked to PNG's $100m carbon trading scandal" (*The Sydney Morning Herald*, September 4, 2009) revealed how a carbon trading scheme in Papua New Guinea thwarted Climate Change Minister Penny Wong's plan to garner support for global carbon trading at the UN conference. Under this scheme, Carbon Planet, an Australian company, issued false carbon certificates to a group of landowners "in order to persuade them to sign over the rights to their forests." It referred to chief executive of the company, Dave Sag, who admitted "that his PNG partner,

Kirk Roberts, had used mocked-up carbon certificates signed up by Mr. Yasasuse as 'props' when negotiating with landowners. But he denied media reports in PNG [that] the certificates were stolen or were intended to mislead." This well-rounded article also attempted to verify the issue with Climate Change Minister Penny Wong, who declined to comment on the matter. However, in the view of the Wilderness Society's Tim King, it was "a tsunami of carbon traders spreading across PNG. Carbon finance and REDD (Reducing Emissions from Deforestation and Degradation) triggered a 'gold rush' mentality." This article sought to convey the gravity of carbon policy-related irregularities by using an activist source.

Shortly after the publication of this article, *The Australian* also published an article on the issue ("It's a market, but in jungle camouflage," *The Australian*, October 23, 2009). This article illustrated the enthusiasm of the financial sector to enter into carbon markets such as REDD. The business sources used in the article wanted to join the carbon market to "make a difference." The article neither included any comment from politicians or government officials about the "controversial" REDD project, nor raised any questions about the controversy surrounding this project. However, it mentioned the "mystery" associated with the carbon credit deal in Carbon Planet without directly alluding to the controversy surrounding this company via any other sources. Analysis of the coverage of REDD indicates that *The Sydney Morning Herald* not only cited more sources, but attempted to cross-check the issues by using government and activist sources. The uncritical presentation of business sources in *The Australian* indicated that business sources were not subjected to the same degree of rigorous scrutiny and verification applied to sources from other areas, such as politics and science (see Ericson et al. 1989; Bacon & Nash 2012). Following Gans (1979, p. 129), perhaps the coverage demonstrated the journalists in *The Australian* perceived the information provided by the business sources as reliable and requiring "the least amount of checking." For Gans, "credibility" is a crucial element of the suitability of sources, although there is dispute over whether judgment of "source credibility" relies on news makers' intuition or rational judgment.

In 2015, business sources were extensively used to argue in favour of the coal industry in Australia on moral grounds, such as addressing "energy poverty" in India. One article, titled "Miners rise to Greens bait on coal for India" (*The Australian*, October 19, 2015), used two business sources and cross-checked them with the then Energy Minister Josh Frydenberg, who described the Australian Greens argument against coal as a "sick joke." As part of the other side of the story, the article used a statement from the Greens Spokesperson, Larrisa Waters, who said: "Burning coal causes local health impacts, with millions of premature deaths from air pollution a year." However, the article again referred to Frydenberg, who said: "I think there is a strong moral case here … More than two billion people today are using wood and dung for their cooking. Now, the World Health Organisation says that this leads to 4.3 million premature deaths." Overall, the story cross-checked with a highly authoritative source (government minister) and strengthened the position of the Mineral Council, but did not seek any comments from the Australian Greens. Thus, the coverage showed a lack of critical scrutiny of the issue of coal mining in Australia. In reference to the opposition Australian Labor Party's proposed energy policy, another report suggested: "Business has big questions" (*The Australian*, July 24, 2015). Here, the journalist argued that the coal industry and the Australian economy faced serious problems from the opposition Labor Party's proposed energy policy, which aimed to address emissions reduction in Australia. In one of the stories, the CEO of the nation's peak business body, Business Council of Australia, Jennifer Westacott, expressed deep concern "about the proposal ... there are also big question marks around the practicality of implementing the policy given the significant disruption to the electricity sector involved" (*The Australian*, July 24, 2015). Interestingly, none of the other three business sources in this article stated that a 50 per cent target was not achievable, although they acknowledged that there could be challenges; for example, more wind turbines (10,000–11,000) meant more sites for renewable projects.

In 2009, the business sources received critical scrutiny at least from one side of the debate, that is, in the pages of The *Sydney Morning Herald*. In 2015, however, the positioning of business sources tended to be more positive in both newspapers. In one article, "Business leader laments 'willful' neglect of science" (*The Sydney Morning Herald*, August 3, 2015), Cameron Clyne, former head of the National Australia Bank, was quoted in a critique of the then Abbott government's renewable policy, which provided reduced support for small solar projects: "climate change poses an 'existential' threat quite unlike any that we have faced before ... The speed of the renewables revolution is 'staggering,' he says, and it would be 'economically reckless' for Australia to remain largely coal-reliant." The article argued that there was strong support from the Australian public for wind and solar energy, as revealed in the survey sponsored by a business body, Future Super (a fossil fuel-free superannuation fund), to investigate the plausibility of Labor's proposed 50 per cent target of renewable energy by 2030 (Hannam 2015). The article used three sources— a government minister, a business source and an expert. The government minister, Greg Hunt, was quoted as saying, "They just pick figures out of the air ... Let's see real detail of their (Labor's) plan—not just for a massive hike in electricity prices on one front, but for a carbon tax on the other." Yet the principal source, the head of Bloomberg New Energy Finance, Kobad Bhavnagri, thought "costs are likely to be manageable," stating: "Our predictions are that just the market naturally—without any new policy—will move to 37 per cent renewable by 2030." Pointing to increasing consumer preference for renewables, he suggested that "the relative cost of renewables versus fossil fuels can be expected to widen over time." However, according to an expert from the Australian National University, Prof Frank Jotzo, co-author of a study on deep decarbonisation: "The most important factor will be bipartisan backing. 'What is needed is a stable, predictable framework,' Jotzo says. 'One side coming out with an ambitious target doesn't guarantee that policy will be achieved.'"

Through the positioning of business, government and activist sources, the articles in both newspapers shifted the framing of the business-dominated climate change issue from contestation or conflict to action between the study periods. The reason behind the shift perhaps lies in the fact that, in 2015, the business community seemed to be divided between those who wanted to continue the production of coal with the introduction of technological support (e.g., CCS), and those who preferred new solutions to emissions reduction through new investment opportunities in the renewable sector. It can be concluded from the above analysis that business actors in 2015 were positively evaluated because of their commitment to emissions reduction efforts in the absence of serious engagement in any debates concerning existing environmental issues (Bacon & Nash 2010). In 2015, while the overall number of business sources increased, the number of business officials as principal sources did not. Despite the strong business implications of the Paris summit, which had a strong investor presence (Benabou et al. 2017), the business sources did not receive significant attention. This observation was partially confirmed by Painter (2016) and his colleagues who analysed COP21 coverage in online news media.

Expert sources

Expert sources, who can be extremely useful in understanding complex climate change issues, are mainly cited in news articles in two ways: first, in relation to the scientists' investigations of climate change, which often contribute enormously to public knowledge; and, second, in response to political statements about climate change, for which the experts provide verification. The role of experts is that of "persons of authority" who either support or challenge various political positions on climate change (Ferree et al. 2002; Tankard 2001, cited in Nisbet 2010). Expert sources played a significant role as "authorised knowers" in the coverage of climate change in both Copenhagen and Paris. In quantitative terms, these expert scientific sources, while fewer in number than the political sources, were more numerous than bureaucratic sources (Appendix 2, Table A2.5). The periods leading up to COP15 and COP 21 were characterised by two particular features of news articles—scientists' and experts' informed opinions and their interpretations in the context of various investigative reports on climate change. In some situations, however, the scientists themselves suddenly became the targets of a systematic campaign that attempted to discredit their findings and assessments of climate change. In November 2009, the Climate Research Unit (CRU) at the University of East Anglia became involved in an email controversy, when an unidentified individual or group hacked the unit's server and illegally published thousands of emails containing scientists' exchanges regarding climate change research on a Russian website. This incident became known as "Climategate."

Knowledge of both the experts' and journalists' responses to such situations is helpful when examining the coverage of what are colloquially called "smear campaigns." Smear campaigns have significant newsworthiness because they revolve around conflict. The two newspapers published stories on scientific leaks or incidents of illegal email exposure. Here, experts attracted significant attention not for their expertise but for the hacked emails that revealed efforts to stave off disagreement between climate scientists. Both internationally and nationally, Climategate was seen as the "worst scientific scandal of our generation" (Booker 2009, cited in Boykoff 2011, p. 37). Even climate activist and *Guardian* journalist George Monbiot called for the resignation of the CRU's director, Professor Phil Jones.

This exposure clearly strengthened the position of climate sceptics, such as the government of Saudi Arabia and media organisations including *The Australian* (McKnight 2010; Manne 2011; Duarte & Yagodin 2012), who sought to interpret the debate surrounding the leaked emails as evidence invalidating climate change expertise. This further polarised discussion of the issue on the eve of the Copenhagen climate summit. The head of the Intergovernmental Panel on Climate Change (IPCC), Dr Rajendra Pachuri, was quoted in relation to the claims about the email leak: "We will certainly go into whole lot, then we will take a position on it. We certainly don't want to brush anything under the carpet" ("Hacked climate emails 'ignored'," *The Australian*, December 5, 2009). The media, however, seemed unconvinced by Pachuri's assurance. The executive director of the Institute of Public Affairs, John Roksham, described the emails as "revealing"; but chose not to comment further. Rather, he expressed his concerns over the local Australian news media's lackadaisical attitude towards this significant issue, attributing it to journalists' "group think mentality" (Schudson 2006). Journalists on the other side of the debate seemed not to accept it. Professor Ove Hoegh-Guldberg, a Marine Science researcher at the University of Queensland, was quoted from an interview:

> Few out-of-context quotes gained by illegally trawling through electronic garbage did not undermine the huge amount of peer-reviewed scientific data on climate change. I think the denialist movement is desperate, given the overwhelming conclusions of the science, that they'll do anything.
>
> ("Climate email mess hits Australia," *The Sydney Morning Herald*, December 5, 2009)

The cross-checking of the views of experts with other experts, including climate believers and climate sceptics, clearly demonstrates the differences in how the selected articles in the two newspapers perceived the issue of email exposure. This was hardly surprising. The articles in *The Australian* portrayed the leak as a severe blow to climate change science and attributed the responsibility for "incorrect data" to the CRU scientists. In the process of verification, which extended to a number of articles immediately after the illegal publication of the leak, this paper not only cited Phil Jones, who argued that the emails were "taken completely out of context," but also John Roskam, a member of a well-known "independent public policy think tank" (Institute of Public Affairs). Thus, one can clearly see how the selection of particular sources and the exclusion of others can contribute to very different interpretations of a highly crucial scientific issue in the articles in *The Australian*.

At the same time, it raises questions about the adequacy of scrutiny by journalists. Following Boykoff (2008, 2011), it can be said that Climategate provided an opportunity to reinforce journalistic norms of balance by offering two contesting viewpoints on this matter. However, the conflict between the scientists and sceptics had a detrimental effect on the outcome of the Copenhagen summit (COP15). What matters here is the way in which experts are "constituted and signified" (Cottle 1998, as cited in Boyce 2006), particularly in the process of cross-checking in which reporters used non-experts (e.g., IPA sources) or other information providers to cross-check expert views. The expert sources were also used to report the impact of climate change, in both 2009 and 2015. For example, in 2009, former Prime Minister Kevin Rudd's call for the opposition to support his government's ETS was based on the premise that rising temperatures across South Australia and Victoria were evidence of climate change. The journalist questioned the association of rising temperatures with climate change, incorporating the viewpoints of Dr Blair Trewin from the National Climate Centre (NCC). The insertion of Dr Trewin's view as an expert source refuted the PM's position to some extent.

> Heatwaves could not be "definitively" linked to climate change. Any individual heatwave like the one we are having at the moment … you can't say definitively it is because of climate change. What we can say is as the overall average temperature increases, and there is a clear increase of average temperatures by 0.8°C in the past century, we would expect the frequency of high extremes to increase and frequency of low extremes to decrease.
>
> ("Heats on to approve carbon plan," *The Australian*, September 28, 2009)

Although Dr Trewin qualified the Prime Minister's claim concerning the connection between rising heat and climate change by using the term "definitive," he did not resile from the fact that climate change was happening. The use of the term "definitive" could also be explained from a science communication perspective, in which the outcome of science was often more "uncertain" than expected from a common-sense point of view (Nelkin 2005; Zehr 2000; Carvalho 2007; Boykoff 2008). In this case, it could be argued that this article focused on "uncertainty" and dramatised the cause of the heatwave by checking the association between heatwaves and climate change with the former head of the NCC, William Kininmonth. Introduced as a climate change sceptic who disapproved of the CSIRO and the BOM report, Kininmonth opined that the prediction of rising temperature was "not going to come true. These (heatwaves) occur randomly, and I don't see see [*sic*] any reason they should come any more often" (*The Australian*, September 28, 2009).

In 2015, expert sources were mainly used to identify economic and security challenges that emerged from the government's and opposition's carbon policies as well as extreme weather events. A case in point here is policy researcher Dr Bjorn Lomborg, who was used extensively to offer his opinion about the Paris conference outcomes and who labelled Paris pledges to reduce global temperature rise by less than 0.05°C by 2030 "wishful thinking." Based on his research paper published in *Global Policy*, he added that the executive secretary of the UN Framework Convention on Climate Change (UNFCC), Christiana Figueres, "entirely misrepresented the world's option" ("Paris pledges 'wishful thinking,' not optimism," *The Australian*, November 10, 2015). The article included statements from French Foreign Minister and COP21 President, Laurent Fabius, and Secretary, Christina Figueres, but did not explicitly cross-check the validity of Dr Lomborg's claim.

Another article, titled "Coal to pay the price of ban on carbon permits" (*The Australian*, August 22, 2015), directly referred to economic modelling by Professor Warwick McKibben, which "shows that under Australia's emissions reduction target of 26 per cent, coalmining could fall by 14.8 per cent without the purchase of international permits. But if international permits are allowed, the fall would be only 7.7 per cent. ... Stronger targets have larger economic impacts." The article mainly argued in favour of Australian business sources' support for purchasing international carbon permits to avoid potential negative impacts on the coal industry and the Australian economy. It also cross-checked with the then Environment Minister, Greg Hunt, and Climate Institute Deputy Chief, Erwin Jackson, in support of the industry. Both sources agreed in principle to "significantly reduce Australia's emissions while still growing the economy." The impact on the economy was the major concern here.

Expert sources were also used to evaluate government policies. The articles, particularly in *The Sydney Morning Herald*, argued that the government's current policy (such as Direct Action) was missing a "critical opportunity" to address public demand for climate action, renewables and emissions reduction—"Backlash looms over climate policy" (*The Sydney Morning Herald*, August 10, 2015). In reference to the Climate Institute research, the article quoted the Chief Executive of the CCA, John Connor, who opined that government actions were "increasingly out of touch with mainstream views." This view was reiterated by another expert from the Energy Change Institute.

In terms of impact, the articles in 2015 witnessed a significant shift to specific climate-related issues (such as bleaching) as opposed to the broader warming debate that prevailed in 2009. On that issue, *The Australian* tended to report the advantages of climate change, for instance citing theoretical biologist Belinda Medlyn's argument that "trees [were] capitalising on abundant carbon dioxide to make up for lack of water" ("Trees reap benefits of climate change," *The Australian,* October 17, 2015). *The Sydney Morning Herald,* in contrast, presented the ongoing impact of climate change: "Mass bleaching threatens reefs" (*The Sydney Morning Herald,* October 9, 2015), and "Diggers face their next battle: extreme weather" (*The Sydney Morning Herald,* October 24, 2015). The latter article made the point that Australia was likely to be challenged by the "unfolding humanitarian disaster" in neighbouring Papua New Guinea. The article used three expert sources who had authored a paper titled "Be prepared: Climate change, security and Australia's defence force." These authors warned that "the politicisation of climate in Australia has left the country's military under prepared," particularly when compared to those in the US and Britain. The expert sources were validated by two bureaucratic sources. The inclusion of defence experts in this article has the potential to shift the news media discourse on climate change to national security issues (see also Olausson & Berglez 2014).

In summary, the newspapers' use of experts shifted between the two periods. While in 2009 the focus was on climate science (informing readers about the possible impact of changing climate as well as publishing leaked documents showing disagreement among scientists), in 2015, the focus was mostly on climate policy and its potential impact on the economy. Experts were positioned in an action frame in which contending parties needed to come together to slow down the adverse impact of climate change. The use of expert sources provides insight into the complex ways in which journalists tried to influence the debate. In quantitative terms, the share of expert sources increased. Arguably, this increase reflects the "increased complexity of a modern 'knowledge society' that increasingly warrants journalists to consult experts in order to interpret and explain climate policy issues and the COP events" (Boyce 2006, p. 890). Some critics, such as Beck (1992), argued that experts were "playing off each other" in a "cacophony of experts" that can be seen as constructive for democracy since it provided the public with an opportunity to become experts themselves. The climate change coverage also intimated that experts had the potential to liberate the public from their unfamiliarity with climate change, but the success of this potential would depend on how journalists used them. Was it merely to maintain the professional norm of objectivity? Or was it, in the name of objectivity, to use them as a smokescreen to avoid acknowledging their "decidedly partisan and political perspective"? (Boyce 2006, p. 890). The selection of information providers showed that "media tend to use 'experts' whose reputation and qualifications add weight to the argument, influence the way events are being interpreted, and set the agenda for future debate" (Rowe et al. 2004, p. 161). Nevertheless, the use of expert sources can be seen as part of a process of "compensatory legitimacy" in which experts are quoted not to unpack complex issues but to confirm the preconceived news frame (Wien 2001; Albaek 2011).

The above analysis demonstrates how the articles in *The Australian* systematically perpetuated doubt through the use of expert sources and verification as well as through interpretation of selected experts' statements. However, climate doubt does not always originate in sources that express scepticism

(Antilla 2005; Boykoff 2007). The point here is the extent to which the articles in *The Australian* consistently represented experts' observations about climate change as an "uncertain puzzle" (Zehr 2000). This position was clearly expressed in an editorial which argued that "science is yet to be convinced" in relation to climate change. It was not simply the selection of sceptical sources, but rather the interpretation of "believers" (scientists) as principal sources that allowed them to promulgate a "sceptical position" that delegitimised climate science and, hence, climate policy. The other side of the debate showed the contestation between the experts on climate science and climate policy. The use and verification showed that the frames were contingent and determined by the purpose of the news makers.

Activist sources

This study found that approximately 10 per cent of the total principal sources in 2009 and 15 per cent in 2015 (Appendix 2, Table A2.5) were activists. The minimal presence of activist sources confirms the marginalisation of non-dominant sources (Schlesinger 1990; Cottle 2002; Lester 2010) and the "fundamental asymmetry implied in the greater power of the media system" (Gamson & Wolfsfeld 1993, p. 115) in mediating environmental politics. Interestingly, the number of activist sources increased between the two study periods. Although activist sources were numerically insignificant, they can at times have the power to reach readers through the "politics of spectacle" (Manning 2001), using available symbolic resources to advance their concerns about climate policies. Both newspapers cited these non-dominant sources to challenge other sources. For example, *The Sydney Morning Herald* used them to directly challenge the industry and government, and *The Australian* did so in relation to the Federal government's proposed ETS and the activists' anti-coal campaign. One of the international investigative articles, titled "Come in spinner" (*The Sydney Morning Herald*, November 7, 2009), presented a clear picture of how the coal lobbies in the US, China and Australia, and the timber lobby in Brazil, attempted to garner public support to resist government policies on carbon pricing. This article juxtaposed two types of activists in relation to the then Rudd government's climate change policy: an industry activist and an environmental activist. It shed light on how a range of industries were trying to squeeze extra financial assistance out of the government. The article stated:

> A Herald analysis of government registers of lobbyists reveals about 120 companies potentially affected by climate change laws employ firms with a total of more than 300 lobbyists … Nearly half the lobbyists working for these firms are former politicians, senior government bureaucrats, or political advisers.

The issues raised in the article were verified by Paul Toni, a climate campaigner for the environmental group World Wildlife Foundation (WWF), who opposed the government's pecuniary measures for industries: "We can't credibly go overseas and say that we will reduce emissions [because] in fact we outsource the whole lot to other countries." Paul Howes, who represented an industry activist group and was at that time Federal Secretary of the Australian Workers Union (AWU), echoed the industry position. Howes privately told his colleagues: "jobs would be lost without the planet being saved." This article held the industry directly responsible for the delay in legislating the ETS locally. However, it was unable to elicit any comment from the relevant executive sources in the industry. It noted its attempt to include industry perspectives from International Power's Australian chief Tony Concannon and TRU energy chief executive Richard McIndoe, and indicated that both declined to be interviewed.

When journalists attempt to scrutinise authorised and accountable sources but cannot do so through these sources, what does this mean for their professional norms, such as balance, or the credibility of their observation as fact (Pan & Kosicki 1993), particularly when such responsibility lies with a particularly influential group? This potentially opens up the use of secondary sources as an important element of analysis. Secondary source materials often allow journalists to provide a good inter-textual connection between the issues raised; they also reinforce the fact that business sources do not readily invite traditional journalistic scrutiny. This may explain the comparative lack of verification of statements in *The Sydney Morning Herald* articles and the slightly better verification track record in *The Australian*. The article that appears below used secondary source material extensively to verify the activist sources in *The Australian*. "Activist" sources promoting climate change at the Copenhagen conference included Leah Wickham, a Fijian Greenpeace activist and delegate at the conference, who was quoted in *The Australian* as saying:

> Fifty years from now, my children will be raising their own families. It is my hope that they will be able to call our beautiful islands [in Fiji] home. It is my hope that our culture and our identity will never be compromised. I'm on the front line of climate change.
>
> ("Strange climate of neglect," *The Australian,* December 12, 2009)

This comment by Wickham invoked an interesting connection between third world claims and first world science. The item verified this Greenpeace activist's claim using the monitoring report of Adelaide-based BOM's National Tidal Centre:

> The tidal centre's equipment shows that the sea level in Fiji has been rising by 5.3mm a year since the equipment was installed in October 1992. The average annual global rise has been about 3mm in recent years; double that of the 20th century. At that rate the sea level around Fiji will be 47cm higher by the start of next century. It is a safe bet that in 2059,

Wickham's children will still be able to call their beautiful islands home.
By then, the sea will probably have risen 26.5cm.

(*The Australian,* December 12, 2009)

While the four sources in this article did not completely agree about the
extent of the sea level rise in Fiji, they did mention the "plight of Kiribati"
(an island nation in the central Pacific) which was experiencing a serious
threat from sea level rise along with an increasing number of "king tides."
Nonetheless, one can see how the article finally associated Wickham's call to
save her island nation with a strategy for asking for $163.5bn in climate aid
from Western industrialised nations, including Australia, which was already
spending $150 million. The article proceeded to raise questions about the
justification for spending Australian taxpayers' money:

How much of that proposed $US150bn would be likely to be spent use-
fully? How much of any cash pumped into the developing nations' climate change kitties would be diverted from other, longer established
projects and programs for supporting those countries? And what of the
environmental problems palpably caused by islanders: the disposable
nappies and drink cans that clog lagoons, the new causeways that desta-
bilise tidal flows around islands, the average in Kiribati of almost six
children a family?

(*The Australian,* December 12, 2009)

These were among the serious charges levelled against the Pacific Island
nations in this article. However, there was no verification from any relevant
authority, nor was there any indication of the context of the sources related to
these questions. In 2015, activists were also used to contest industry initiatives.
For example: "Environmental groups 'devastated' as BHP starts Borneo coal"
(*The Sydney Morning Herald,* July 23, 2015) described the environmental
campaigners' frustration with the commencement of coal production by BHP
Billiton at Haju and IndoMet mines in Indonesian Borneo in the province of
Central Kalimantan, an area of rich bio-diversity and home to a significant
orangutan population. When contacted, the BHP spokeswoman said:

BHPB has approached the development of IndoMet in a cautious
manner. In doing so, we have adhered to our world-class environmental
and governance standards and practices ... Our approach to operating in
a sustainable manner can make a positive contribution to the region and
we will only proceed with a development that is consistent with our
sustainability commitment.

During both study periods, the industry sources either declined to comment
or denied the potential for any "environmental despoliation" to result from
their commercial activities and undermine environmental protection. These

denials from industry and business sources can be seen as evidence of their reluctance to engage with environmental concerns (Bacon & Nash 2010). In 2015, *The Australian* continued to focus on activist sources but confined their scrutiny to industrial matters in Australia. For example, one article ("Risk to reef new front in coal campaign," November 10, 2015) demonstrated a direct contestation between the activist from the Australian Conservation Foundation (ACF) and the government over the latter's failure to take into consideration "climate pollution" of the Great Barrier Reef from the proposed Adani mine. Referring to the legal challenge in the Federal court, ACF President Geoff Cousins said:

> This action is historic, it's the first case that has sought to test the Environment Minister's World Heritage obligations as they relate to the climate change impacts on the reef caused by pollution from burning a mine's coal … If this case is successful it will strengthen climate change considerations and World Heritage protection in Australian law.

From the perspective of the Adani sources, environmental groups were "seeking endless delay and endlessly abusing process." The article described the litigation by the ACF as "lawfare" that prompted a federal government move to limit "the ability of green groups to frustrate major projects in the courts." This conflict was triggered by government decisions on national climate policy and environmental approval for Adani. This environmental challenge escalated from a regional to a national issue when

> In August, then prime minister Tony Abbott said he wanted to repeal section 487 (2) of the EPBC Act, which had allowed green groups to challenge developments on the grounds they affected the environment or an endangered species even if the groups had no direct interest in the project.
>
> (*The Australian*, November 10, 2015)

> Mr Cousins denied the latest challenge was part of a conservationist attempt to frustrate Adani with delays and force it to abandon the project. "This is not 'lawfare'; it's about the concept of fair law in being able to contest a decision made by governments that is at the heart of a democracy," he said.
>
> (Ibid.)

In relation to the process of source verification, it can be deduced that the journalists were more concerned with substantiating the position adopted in these articles. This is why, in 2009, when journalists held industries responsible, there was active use of industry sources whereas, in 2015, journalists labelled activists' legal action against a mining company as "lawfare" and sought verification from activist sources. Overall, through these verifications, journalists framed the issue as a contestation between activists and industry, invoking the environment versus development debate in the public sphere. It can be argued that, although the number of sources increased between these two periods, this did not result in access for activist sources in terms of news media privilege. As critics (Ericson et al. 1989; Manning 2001) have argued, sometimes activist-generated events secure coverage but not access. However, in this example, the activists may not have had privileged access to news, but they were at least capable of challenging businesses through the legal system (i.e., Cousin's legal challenge to Adani and government).

Conclusion

The analysis of the process of cross-checking sources confirms that political authorities, including the government and opposition, as well as experts, were the dominant actors in the coverage of climate change during the two periods, while business, activists and citizens were less influential sources. Overall, the analyses of sources demonstrated that political sources were the most widely used (Bennett 2010; Kim & Weaver 2003; Dimitrova & Strömbäck 2009) but the least verified. The substantial presence of non-verified sources can be explained by the characteristics of the coverage, which generally featured event-based reports about the summits and climate change policies. In these scenarios, journalists often lack the ability to verify statements due to tight daily deadlines (Tumber 2010; Tuchman 1978). The coverage also shows that "written prescriptions of verification routines" are yet to be applied in daily events-based reporting (Shapiro et al. 2013, p. 659). It can also be argued that the journalists selected sources that had the greatest access to the "hierarchy of credibility" (Becker 1967) and were less likely to seek verification because of trustworthiness attributed to powerful sources, which led news workers to compromise cross-checking for pragmatic reasons. However, the fact that political sources were verified less often but expert sources were the most frequently cross-checked, raises questions about the practice. Perhaps this tendency is associated with the increasing number of expert sources (see Boyce 2006). Although these experts are accessing news media with their own agendas (Cottle 2000), at times they were used by journalists to frame the climate argument.

The analysis also indicates a very infrequent but convenient use of activist sources in the coverage of climate change. Articles in *The Sydney Morning Herald* challenged the private sector using activist sources to scrutinise the government and the business sector. *The Australian* used activists to challenge the government's proposed climate change policy. While the under-representation of activists is obvious, what is more crucial here is the way in which both newspapers amplified their respective positions on climate change policy by citing an activist as an auxiliary source. Such citing functioned as a subsidiary to other mainstream powerful sources, such as politicians and business officials. Analyses of the sources in *The Australian* revealed that the characterisation of climate change was quite clear. From a political source perspective, it demonstrated a contestation between rich and poor countries that were unwilling to commit to emissions reduction. Having agreed with the significance of the climate deal, *The Australian* went against the premise of climate change policy for two main reasons. First, it argued that the introduction of a local climate change policy would be devastating for the economy and would result in "job losses." The development of alternative energy resources (nuclear energy) was its priority. Second, it questioned the projections around climate change by interpreting the experts' quotes in a way that contested the relevant issues, at both the political and scientific levels.

In contrast to *The Australian*'s stance, the articles in *The Sydney Morning Herald* framed the political issues surrounding climate change—particularly the Copenhagen summit—positively. These articles portrayed the representation as a contest between politicians across the world. However, in relation to local climate change issues, the articles were quite critical of both the government and businesses on two main grounds. First, they claimed that the absence of adequate government scrutiny was rendering various climate change policies ineffective. Second, they asserted that various activities of the private sector were hindering the government's effort to introduce a viable climate change policy.

The delineation of climate change issues in terms of journalistic content can also be understood through the visibility of certain sources and the invisibility of others, in both quantitative and qualitative senses. Quantitative invisibility suggests the comparative absence of some sources, such as activists, while qualitative invisibility refers to a source that is cited expediently despite having a low numerical presence. The articles in the two newspapers used experts (more visible) and activists (less visible) as sources in an action frame that emphasised their positive and negative orientations towards environmental policy-related issues. These positions were clearly evident in their coverage of the impact of climate change. Articles in *The Sydney Morning Herald* gave prominence to the experts' observations in news that bolstered the broadsheet's positive orientation towards the climate change issue. Conversely, articles in *The Australian* adopted two specific strategies to endorse their negative orientation towards the issue: first, the selection of certain sources (e.g., farmers) to refute evidence of climate change provided by the experts; and, second, the selection of certain quotes and interpretations that enabled the newspaper to raise doubt about anthropogenic climate change. Both publications used less visible sources—activists—to legitimise their positive and negative orientations towards climate change issues. In this process of legitimisation, the articles in *The Australian* attempted to influence issues in the political field by challenging the political sources, while the articles in *The Sydney Morning Herald* did so by scrutinising the business sources. The articles in *The Australian* tried to influence issues in the political field by using business interests to challenge political sources. The broadsheet explicitly associated the Australian government's proposed climate change policy with potentially negative economic consequences, such as job losses in different sectors. The analysis of business, expert and activist sources clearly demonstrates how the articles in this newspaper attempted to perpetuate reservations about climate change by questioning both climate science and relevant public policy. Nonetheless, at times, the newspaper hinted that business sources could favour the government's climate change policy; for example, business communities' demand for "clean coal" and the "nuclear energy" option to reduce the country's reliance on fossil fuels.

Thus, as suggested previously, it is not only the selection of sources, but also the presentation and interpretation of sources through the process of framing that enables news organisations to "simultaneously invoke and apply norms" as well as to "define them" (Tuchman 1978, p. 184). This process of defining certain issues, while crucial, is to some degree contingent rather than predetermined. As a result, and as the above discussion shows, there is deviation in the continuum of the anti- and pro-climate change significations.

References

Albaek, E. 2011, 'The interaction between experts and journalists in news journalism', *Journalism*, vol. 12, no. 3, pp. 335–348.

Antilla, L. 2005, 'Climate of skepticism: US newspaper coverage of the science of climate change', *Global Environmental Change*, vol. 15, no. 4, pp. 338–352.

Arup, T. & Kenny, M. 2015, 'Paris UN climate conference 2015: Australia to double clean energy research spending', *Sydney Morning Herald*, accessed December 14, 2018, available: https://www.smh.com.au/environment/climate-change/paris-un-climate-confer ence-2015-australia-to-double-clean-energy-research-spending-20151130-glb0wi.html.

Bacon, W. & Nash, C. 2010, 'Playing the media game: the relative (in)visibility of coal industry interests in media reporting of coal as a climate change issue in Australia', paper presented at the pre-ECREA conference: Communicating climate change II, Institute of Journalism and Communication, University of Hamburg, Hamburg, Germany.

Bacon, W. & Nash, C. 2012, 'Playing the media game', *Journalism Studies*, vol. 19, no. 3, pp. 1–16.

Beck, U. 1992, *Risk society: Towards a new modernity*, translated by M. Ritter, Sage, London.

Beck, U. 2009, 'Critical theory of world risk society: A cosmopolitan vision', *Constellations*, vol. 16, no. 1, pp. 3–22.

Becker, H. S. 1967, 'Whose side are we on?' *Social Problems*, vol. 14, no. 3, pp. 239–247.

Benabou, S., Moussu, N. & Müeller, B. 2017, 'The business voice at COP21: The quandaries of a global political ambition', in S. C. Aykut, J. Foyer & E. Morena (eds) *Globalising the climate: COP21 and the climatisation of global debates*, Routledge, New York, pp. 57–74.

Bennett, W. L. 2003, 'Communicating global activism: Strengths and vulnerabilities of networked politics', *Information, Communication & Society*, vol. 6, no. 2, pp. 143–168.

Bennett, W. L. 2010, 'The Press, Power and Public Accountability', in S. Allan (ed.) *The Routledge Companion to News and Journalism*, Routledge, London, pp. 105-115.

Berkowitz, D. A. 2009, 'Reporters and their sources', in K. Wahl-Jorgensen & T. Hanitzch (eds) *The handbook of journalism studies*, Routledge, New York, pp. 102–115.

Booker, C. 2009, 'Climate change: This is the worst scientific scandal of our generation', *The Telegraph*, November 28, accessed May 23, 2016 available: https://www.telegraph.co.uk/comment/columnists/christopherbooker/6679082/Climate-change-this-is-the-worst-scientific-scandal-of-our-generation.html.

Boyce, T. 2006, 'Journalism and expertise', *Journalism Studies*, vol. 7, no. 6, pp. 889–906.

Boykoff, M. T. 2007, 'Flogging a dead norm? Newspaper coverage of anthropogenic climate change in the United States and United Kingdom from 2003–2006', *Area*, vol. 39, no. 4, pp. 470–481.

Boykoff, M. 2008, 'Media and scientific communication: A case of climate change', in D. G. E. Liverman, C. P. G. Pereira & B. Marker (eds) *Communicating environmental geoscience*, Geological Society Special Publications, London, vol. 305, pp. 11–18.

Boykoff, M.T. 2011, *Who speaks for the climate? Making sense of media reporting on climate change*, Cambridge University Press, Cambridge.

Broersma, M. 2010, 'The unbearable limitations of journalism: On press critique and journalism's claim to truth', *International Communication Gazette*, vol. 72, no. 1, pp. 21–33.

Carvalho, A. 2007, 'Ideological cultures and media discourses on scientific knowledge: Re-reading news and climate change', *Public Understanding of Science*, vol. 16, no. 2, pp. 223–243.

Cottle, S. 1998, 'Ulrich Beck, "risk society" and the media: A catastrophic view?' *European Journal of Communication*, vol. 13, no. 1, pp. 5–32.

Cottle, S. 2000, 'Rethinking news access', *Journalism Studies*, vol. 1, no. 3, pp. 427–448.

Cottle, S. 2002, *News, public relations and power*, Sage, London.

Cottle, S. 2009, *Global crisis reporting*, Open University Press, Maidenhead.

Diekerhof, E. & Bakker, P. 2012, 'To check or not to check: An exploratory study on source checking by Dutch journalists', *Journal of Applied Journalism & Media Studies*, vol. 1, no. 2, pp. 241–253.

Dimitrova, D. & Strömbäck, J. 2009, 'Look who's talking? Use of sources in newspaper coverage in Sweden and the United States', *Journalism Practice*, vol. 3, no. 1, pp. 75–91.

Duarte, K. & Yagodin, D. 2012, 'Scientific leaks', in E. Eide & R. Kunelius (eds) *Media meets climate: The global challenge for journalism*, NORDICOM, Göteborg, Sweden, pp. 163–178.

Eide, E., Kunelius, R. & Kumpu, V. 2010, *Global climate, local journalisms: A transnational study of how media make sense of climate summit*, Projekt Verlag, Bochum, Germany.

Ericson, R. V., Baranek, P. M. & Chan, J. B. L. 1989, *Negotiating control: A study of news sources*, Open University Press, Milton Keynes.

Ericson, R. V., Baranek, P. M. & Chan, J. B. L., 1991, *Representing order: Crime, law, and justice in the news media*, Open University Press, Buckingham.

Ferree, M. M., Gamson, W. A., Gerhards, J. & Rucht, J., 2002, *Shaping abortion discourse: Democracy and public sphere in the Germany and the United States*, Cambridge University Press, New York.

Gamson, W. A. & Modigliani, A. 1987, 'The changing culture of affirmative action', in P. Burstein (ed.) *Equal employment opportunity: Labor market discrimination and public policy*, Aldine de Gruyter, New York, pp. 137–177.

Gamson, W.A. & Wolfsfeld, G. 1993, 'Movements and media as interacting systems', *The Annals of the American Academy of Political and Social Science*, vol. 528, no. 1, pp. 114–125.

Gans, H. J. 1979, *Deciding what's news*, Pantheon Books, New York.

Gitlin, T. 2003, *The whole world is watching: Mass media in the making and unmaking of the new left*, University of California Press, Berkeley, CA.

Hannam, P. 2015, 'Australia's large-scale renewable investment dives in 2014', *The Sydney Morning Herald*, January 12, p. 12.

Holmes, D. 2015, 'A Paris summit for climate and peace?', *The Conversation*, accessed June 16, 2016, available: https://theconversation.com/a-paris-summit-for-climate-a nd-peace-51393.

Iyengar, S. 1996, 'Framing responsibility for political issues', *The Annals of the American Academy of Political and Social Science*, vol. 546, no. 1, pp. 59–70.

Jönsson, A. M. 2011, 'Framing environmental risks in the Baltic Sea: A news media analysis', *Ambio*, vol. 40, no. 2, pp. 121–132.

Kim, T. S. & Weaver, H. D. 2003, 'Reporting on globalisation: A comparative analysis of sourcing patterns in five countries' newspapers', *Gazette: The International Journal for Communication Studies*, vol. 65, no. 2, pp. 121–144.

Kovach, B. & Rosenstiel, T. 2007, *The elements of journalism: What newspeople should know and public should expect*, Three Rivers Press, New York.

Lawrence, G. R. 2000, 'Game-framing the issues: Tracking the strategy frame in public policy news', *Political Communication*, vol. 17, no. 2, pp. 93–114.

Lester, L. 2010, *Media and environment: Conflict, politics and the news*, Polity, Cambridge.

Manne, R. 2011, 'Bad news: Murdoch's Australian and the shaping of the nation', *Quarterly Essay*, vol. 43, pp. 1–119.

Manning, P. 2001, *News and news sources: A critical introduction*, Sage, London.

McCombs, M. 2005, 'A look at agenda setting: Past, present and future', *Journalism Studies*, vol. 6, no. 4, pp. 543–557.

McDonald, M. 2015, 'Australian foreign policy under the Abbott government: Foreign policy as domestic politics?', *Australian Journal of International Affairs*, vol. 69, no. 6, pp. 651–669.

McKnight, D. 2005, 'Murdoch and the culture war', in R. Manne (ed.) *Do not disturb: Is the media failing Australia?*, Black Inc., Melbourne, Vic., pp. 53–74.

McKnight, D. 2010, 'A change in the climate? The journalism of opinion at News Corporation', *Journalism*, vol. 11, no. 6, pp. 693–706.

Nelkin, D. 2005, *Selling science: How the press covers science and technology*, W. H. Freeman, New York.

Nisbet, C. M. 2010, 'Knowledge into action: Framing the debates over climate change and poverty', in P. D'Angelo & J. A. Kuypers (eds) *Doing news framing analysis: Empirical and theoretical perspectives*, Routledge, New York, pp. 43–83.

Okereke, C. & Coventry, P. 2016, 'Climate justice and the international regime: Before, during, and after Paris', *Wiley Interdisciplinary Reviews: Climate Change*, vol. 7, no. 6, pp. 834–851.

Olausson, U. & Berglez, P. 2014, 'Media and climate change: Four long-standing research challenges revisited', *Environmental Communication*, vol. 8, no. 2, pp. 249–265.

Painter, J. 2016, 'Disaster, uncertainty, opportunity or risk? Key messages from the television coverage of the IPCC's 2013/2014 reports', *METODE Science Studies Journal*, no. 6, pp. 81–87.

Pan, Z. & Kosicki, G. 1993, 'Framing analysis: An approach to news discourse', *Political Communication*, vol. 10, no. 1, pp. 55–75.

Park, E. R. 1940, 'News as a form of knowledge: A chapter in the sociology of knowledge', *American Journal of Sociology*, vol. 45, no. 5, pp. 669–686.

Peters, C. & Broersma, M. 2017, 'The rhetorical illusions of news', in C. Peters & M. Broersma (eds) *Rethinking journalism again: Societal role and public relevance in a digital age*, Routledge, London, pp. 188–204.

Pham, B. D. & Nash, C. 2017, 'Climate change in Vietnam: Relations between the government and the media in the period 2000–2013', *Pacific Journalism Review: Te Koakoa*, vol. 23, no. 1, pp. 96–111.

Roosvall, A. 2017, 'Journalism, climate change, justice and solidarity: Editorializing the IPCC AR5', in E.Eide R., Kunelius & K. Kumpu (eds) *Global climate, local journalism: A transnational study of how media makes sense of climate summits*, Projekt Verlag, Freiberg, Germany, pp. 129–150.

Rowe, R., Tibury, F., Rapley, M. & O'Farrell, I. 2004, 'About a year before the breakdown I was having symptoms: Sadness, pathology and the Australian newspaper media', in C. Seale (ed.) *Health and the media*, Blackwell, Oxford, pp. 160–175.

Schlesinger, P. 1990, 'Rethinking the sociology of journalism: Source strategies and the limits of media-centrism', in M. Ferguson (ed.) *Public communication: The new imperatives*, Sage, London, pp. 61–83.

Schudson, M. 2006, 'The trouble with experts—And why democracies need them', *Theory and Society*, vol. 35, no. 5–6, pp. 491–506.

Shapiro, I., Brin, C., Bédard-Brûlé, I. & Mychajlowycz, K. 2013, 'Verification as a strategic ritual: How journalists retrospectively describe processes for ensuring accuracy,' *Journalism Practice*, vol. 7, no. 6, pp. 657–673.

Shehata, A. 2007, 'Facing the Mohammad cartoons: Official dominance and event-driven news in Swedish and American elite press', *Harvard International Journal of Press/Politics*, vol. 12, no. 4, pp. 131–153.

Sreberny, A. & Paterson, C. (eds), 2004, *International news in the twenty-first century*, John Libbey Publishing, Eastleigh, UK.

Strömbäck, J. & Nord, L. W. 2006, 'Do politicians lead the tango? A study of the relationship between Swedish journalists and their political sources in the context of election campaigns', *European Journal of Communication*, vol. 21, no. 2, pp. 147–164.

Swain, B. M. & Robertson, J. M. 1995, 'The Washington Post and the Woodward problem', *Newspaper Research Journal*, vol. 16, no. 1, pp. 2–20.

Tankard, J. 2001, 'The empirical approach to the study of media framing', in S. Reese, O. H. Gandy & A. Grant (eds) *Framing public life: Perspectives on media and our understanding of the social world*, Lawrence Erlbaum, Mahwah, NJ, pp. 95–106.

Tankard, J., Hendrickson, L., Silberman, J., Bliss, K. & Ghanem, S. 1991, 'Media frames: Approaches to conceptualization and measurement', paper presented at the Association for Education in Journalism and Mass Communication, Communication Theory and Methodology Division, Boston, MA.

Tuchman, G. 1978, *Making news: A study in the construction of reality*, Free Press, New York.

Tumber, H. 2010, 'Journalists and war crimes', in S. Allan (ed.) *The Routledge companion to news and journalism*, Routledge, New York, pp. 533–541.

van Dijk, T. 1991, *Racism and the press*, Routledge, London.

Walsh, B. 2016, 'Clean energy is worth trillions, John Kerry says', *Huffington Post*, April 6, accessed April 12, 2019, available: https://www.huffingtonpost.com.au/entry/clean-energy-economic-opportunity_n_5703cc2ce4b0daf53af0dc08?ec_carp=59963879427119111108.

Ward, S. J. 2018, *Ethical journalism in a populist age: The democratically engaged journalist*, Rowman & Littlefield, Lanham, MD.

WienC. 2001, 'Journalisters brug af ekspertkilder i danske aviser (The reference to expert sources in Danish newspapers)', paper presented at the Nordmediakonferens, Reykjavik, August 11–13.

Wolfsfeld, G. & Sheafer, T. 2006, 'Competing actors and the construction of political news: The contest over waves in Israel', *Political Communication*, vol. 23, no. 3, pp. 333–354.

Zehr, S. 2000, 'Public representations of scientific uncertainty about global climate change', *Public Understanding of Science*, vol. 9, no. 2, pp. 85–103.

Zelizer, B. & Allan, S. 2010, *Keywords in news and journalism studies*, Open University Press, New York.

6 Climate change news and sources in Bangladesh

This chapter compares the various positions that journalists attribute to different sources to examine the impact of these attributions on the verification of source statements. This verification process requires journalists to pit one source's perspectives against those of other sources or a countervailing response to test the veracity of a claim or an argument. The process involves the sources providing relevant information and journalists deciding what information and which sources to include in the news stories. Because of the nature of this exchange between sources and journalists, the power to determine the interpretation that ultimately prevails over others and influences the social reality of climate change is always at stake (Beck 2010; Broersma et al. 2013, p. 388). The news media's "hypothesis-testing features" (Park 1940; Pan & Kosicki 1993) render them a privileged site for such contestation among different sources (Beck 1992; Cottle 1998). As the pre-eminent agent of this site, journalists have the potential to influence debates surrounding any public issue, including climate change.

The chapter describes various sources and the verification of assertions made by these sources in the news articles pertaining to climate change in *The Daily Star* and *Prothom Alo* in 2009 and 2015. During these two periods the news coverage was heavily dominated by political sources ($n = 302$) followed by bureaucrat ($n = 202$), expert ($n = 122$), activist ($n = 97$), business ($n = 33$) and citizen ($n = 37$) sources. The use of these sources allows exploration of how the journalists made sense of the environmental issues through the use of sources. This exploration facilitates the identification of dominant meanings that might emerge from such representations, especially if there are any differences or similarities between the coverage of these issues in the two periods. Since the focus is on the journalists' use of sources "through the choice of emphasis and inclusion or exclusion of information" (Anderson 2014, p. 43) it is necessary to explore the significance of the presence of certain sources and absence of others. The discussion begins with an analysis of various principal sources—politicians, bureaucrats, activists, experts and citizens. The statements provided by the principal and other sources and subsequent use of these statements in the articles were examined to explicate the functional relations between the sources to build the intended or preferred meaning. The principal sources (see Appendix 2, Table A2.8) were defined earlier as the most influential source/s who made assertions either in support of or against the main theme of the selected article. Analysis of these sources sheds light on the complexities of the coverage of climate change as both a global and local environmental threat. In addition, the analysis of the use of sources also clarifies the contestation between various sources in the climate change news that is positioned at the intersection of the local and the global, providing a larger picture of the newspapers' responses to this complex environmental issue.

The coverage is examined in regard to the use of principal source and verification of source statements, which allows news workers to build certain frames over others to make climate change news appear as legitimate and common sense (Anderson 2014). The analysis identified the ways in which journalists framed their sources. The sources and principal sources that were used to verify or cross-check statements or excerpts in the selected articles were identified as described in Chapter 3 (Appendix 2, Tables A2.8 and A2.9). The process of verification is a defining characteristic of the practice of journalism (Machill & Beiler 2009; Kovach & Rosenstiel 2007, p. 667). For the purpose of the current analysis, following Shapiro et al. (2013), verification is seen as an integral part of the news reporting process that is built on the effective use of source excerpts "step by step and looping back in upon itself" (p. 667) that sustain as well as address readers' curiosity.

Political sources: A matter of a fair share

The analysis of the coverage of the Copenhagen conference revealed the intricacies of politicians' representations, highlighting the difference between various leaders' views of global climate change policy. In the main, the difference related to the varied degrees of political will in addressing the climate policy issues. An example of these intricacies was found in a *Prothom Alo* article: "No agreement so far: World leaders may sign political declaration, waiting for Copenhagen Declaration" (December 19, 2009). This article appeared on the front page, accompanied by an image of some environmental activists shaving their heads in protest against the looming uncertainty of the conference outcome which, they claimed, was due to hesitation on the part of world leaders. The indecision of the leaders and its impact on the Small Island Nations were the focus of this article. This sentiment was also echoed in US President Barack Obama's speech when he said: "The future of the vulnerable small nations, such as Bangladesh, will be more endangered if all the countries across the world are unable to reach an agreement to prevent climate change" (*Prothom Alo*, December 19, 2009). The article also used an excerpt from Indian Prime Minister Manmohan Singh's speech as a sign of India's cautious enthusiasm for a negotiated outcome. For him: "This agreement will be able to demonstrate a profound respect to the peoples' movement on climate change around the world" (*Prothom Alo*, December 19, 2009). These two quotations, one from then US President and the other from the Indian Prime Minister, revealed little about the complexities involved in the negotiation process between the US and the BASIC countries (Brazil, South Africa, India and China), but they did exemplify the vulnerability of Bangladesh to climate change and the country's expectations of the Copenhagen summit.

From this explication, it becomes evident that the article has strategically assigned responsibility to the leaders of "other countries" for their inability to reach an agreement. The ensuing lack of consensus engendered subtle tensions on two fronts: first, between the leaders of the developed and developing countries on the question of emissions reduction (e.g., the US vs China); and, second, between the polluting rich and the suffering poor countries over the question of climate aid (e.g., the European countries vs Bangladesh and Small Island Nations). The following article makes this tension explicit: "Draft climate accord seems inadequate" (*The Daily Star*, December 19, 2009). As the headline suggests, the article anticipated the summit outcome. The journalist used statements from one expert and two political sources, verifying different political statements using the expert source. For example, Dr Saleemul Huq, chief of the climate change cell of the London-based International Institute for Environment and Development (IIED), was quoted as saying that the outcomes were totally "inadequate" for Bangladesh. In this context, the article used terms of low modality, such as "seems" and "inadequate." Although the basis of the headline and introduction consisted of a single word quoted from Dr Huq, the article did not elaborate further on the said "inadequacy." The subsequent information provided by the reporter elaborated somewhat on the "draft text," which failed to commit to either legally binding carbon cuts or to any climate adaptation funds for vulnerable countries, such as Bangladesh. By using this comment in the "enmeshing of fact and source" (Tuchman 1978), the journalist justified it and, thus, highlighted the inefficacy of the potential outcome of the negotiations.

Here, the selection of Dr Huq's comment was particularly significant because Bangladesh's ruling politicians seemed to have decided not to make any negative comment about the progress of the conference negotiations. However, regarding adaptation needs, it expounded on Bangladeshi Prime Minister Sheikh Hasina's call for the inclusion of "climate refugees" in any agreement. This clause would obligate the developed countries to accept Bangladeshi citizens as refugees on the grounds of their vulnerability due to extreme exposure to climate change. She also indicated her willingness to reduce carbon emissions if Bangladesh received "technological and financial help from developed nations." Her statement was corroborated by the country's Environment Minister, Hasan Mahmud, who asserted: "None [world leaders] differed with the Prime Minister's demand." However, Mahmud acknowledged that "none of them announced anything concrete to save the world," a comment echoed in US President Barak Obama's speech in which he stated that "there must be financing that helps developing nations."

The willingness to reduce carbon gained traction during the Paris summit, as indicated in the headline of an article: "Hundred countries united to keep temperature rise below 1.5°C" (*Prothom Alo*, December 2, 2015). This article referred to the commitment made at the Climate Vulnerable Forum (CVF), a side event of the conference, which received support for the temperature containment goal agreed at the main COP21 forum. In this article, the then Environment and Forest Minister of Bangladesh, Anwar Hossain, noted: "As a climate vulnerable country, Bangladesh had already reached at a dangerous level." On the other hand, Costa Rica's Foreign Minister highlighted his country's success in using renewable energy as the main source of electricity generation and was keen to use this success as a model for reducing the current level of emissions: "Many people thought carbon emissions reduction could harm the development, but Costa Rica has proved that it is possible to simultaneously achieve emissions reduction and development." The CVF's emphasis on keeping the temperature rise below 1.5°C was positively verified by the above-cited expert, Dr Saleemul Huq, who noted that the UN had already acknowledged the CVF's demand and stated that, if the forum members were united, they could execute a binding agreement from Paris. From these comments, it can be deduced that the developing nations' main focus in 2015 was on unity to achieve the carbon reduction target.

France, the host country in 2015, was also eager to reach an agreement. One article referred to the French Foreign Minister, Lauren Fabius, in the headline: "France wants momentum in agreement" (*Prothom Alo*, December 3, 2015). While developed industrialised countries showed their commitment to reducing carbon emissions, Bangladesh's Environment and Forest Secretary, Kamalunddin Ahmed, presented his country's commitment to a 15 per cent carbon emissions reduction by 2020 through the Intended Nationally Determined Contributions (INDC). Another Bangladeshi negotiator, Prof. Mizan R. A. Khan, referred to the CVF report titled "Fair shares: A civil society equity review of INDC," stating that the commitments to emissions reduction made by the main carbon-emitting countries, such as China and the US, were deplorable. Thus, through the use of speeches and comments by national and world leaders, the journalist assigned responsibility to "other countries" for the dissension that was threatening the negotiation process, a dissension that left "endangered" small nations, such as Bangladesh, without any significant financial or material support to cope with climate change. Here, the important matter is not the extent of the presence of political sources, but the variation in the use of these sources during the two study periods. While journalists in 2009 used expert sources to demonstrate contestations among various world leaders and to verify the statements made by these political leaders, in 2015, they positioned these experts to reach a consensus or meaningful agreement of the political leaders.

In 2009, political sources were high in number but less frequently verified, while the opposite was the case in 2015. This difference can be explained by a shift in the news value. In 2009, the Bangladeshi delegation was led by Prime Minister Sheikh Hasina; in 2015, it was led by the Forest and Environment Minister Anwar Hossain. This may have resulted in less coverage and use of political sources but increased verification (Appendix 2, Table A2.9). In both periods, the politicians were validated by the expert sources. However, the use of experts' expertise (van Dijk 1991; Boyce 2006) in the positive verification of politicians' statements in both periods demonstrated that the journalists were inclined to fortify Bangladesh's political position and set the agenda for future debates. In 2009, the expectation was that the summit would reflect on the issues of climate aid and climate refugees. This expectation was maintained in 2015, but the articles were subtler and more rational in highlighting the Bangladesh government's policy initiatives to tackle climate change through the emissions reduction commitment made in the country's INDC.

This endorsement by the two newspapers is consistent with what some critics have described as a pattern of relationship between journalism and the nation state in which journalists intervene in the mediation of issues in the best interests of the country (Anderson 1991). In this case, the interests of Bangladesh were similar to those of these newspapers. On the eve of the Paris conference, government ministers explicitly stated their expectation: "Mass media should side with the government in climate conference" (*Prothom Alo*, November 27, 2015). The ministers called on the mass media to support Bangladesh's demands in the Paris conference in a more robust way.

It has been argued (Nassanga et al. 2017) that this type of coverage relates to the idea of development journalism, in which government authorities ask for the media's uncritical support in advancing national interests. While this argument seems to hold true in this instance, it is also important to note that journalists are not always able to adequately verify all the articles they publish, especially those based on events. However, it is also significant to note that, in both 2009 and 2015, politicians were less often verified as sources compared to others. This can be explained by the idea of "hierarchies of credibility," in which political sources enjoy a trusting relationship with journalists because they are often useful as news sources. This puts them in a strong position in regard to their relationship with journalists and, conversely, weakens the news workers' ability to critically scrutinise them as sources. While both newspapers relied heavily on political sources in the "hierarchies of credibility," perhaps because the assertions of political and official sources can be reproduced in news without much effort to ensure their veracity, the publications also focused on bureaucratic sources to frame climate change news.

Bureaucratic sources: From infuriation to action

In both newspapers, the number of articles dominated by bureaucrats was significant in comparison to other influential sources, such as experts (Appendix 2, Table A2.7). The bureaucrats were largely consigned to the issue of climate change as a matter of contestation (Neuman et al. 1992; Semetko & Valkenburg 2000) between various international bureaucrats and politicians over the predicaments of the Copenhagen summit. However, in the lead-up to the Paris summit, bureaucratic sources were used to expound the vulnerability of Bangladesh, as a deltaic land, to the onset of climate change. This usage was particularly prominent in 2015. In 2009, the ways in which bureaucrats were used to render the politicians responsible for the stalemate in negotiations indicated an invisible conflict between the two groups. This use of international bureaucrats was further strengthened by the perspectives of activist sources, which is discussed later in this chapter. During the Copenhagen summit in 2009, the coverage succinctly summarised the developing countries' concerns about the actions of the developed world. This was achieved by using bureaucratic and other sources, including politicians and activists. For example: "Danish text leak sparks debate over talk's success" (*The Daily Star*, December 10, 2009) and "Leaked draft: Conference centre in chaos" (*Prothom Alo*, December 10, 2009). Both articles highlighted the revelations in the Danish Draft titled "Adoption of the Copenhagen Agreement." They used two different bureaucratic sources that were subsequently verified by experts, politicians and Non-Governmental Organisation (NGO) sources. While *Prothom Alo* presented this issue in a feature article, *The Daily Star* published a hard news story directly pointing to the looming uncertainty of the summit's success. This was echoed in a statement made by the G77 Chief Negotiator, Lumumba Stanislaus Di-Aping: "I would like to say on behalf of G77 that this is a very serious development. It's a major violation that threatens the success of the Copenhagen negotiations" (*The Daily Star*, December 10, 2009). Referring to the leaked draft, the article elaborated on how the draft by-passed the spirit of the Bali Action Plan. This Plan recommended a legally binding agreement and adequate assistance for developing countries. The assistance would depend on appropriate increases in mitigation and adaptation efforts by developing countries; in other words, any money to help poor countries adapt to climate change would depend on their actions (*The Daily Star*, December 10, 2009).

The inclusion of conditions for climate aid infuriated developing countries. *Prothom Alo* described the draft as a "very dangerous document for the developing countries" (December 10, 2009). The leaders of the developing countries expressed their exasperation and received support from NGOs, including the World Wildlife Foundation (WWF) and Oxfam International. What is important to note here is not only the prominence of the voices, but also the strategic use of certain sources: that is, in the process of vigilance, journalists verified the bureaucratic assertions by incorporating activist and expert sources who were comparatively less visible. As Antonio Hill, the Climate Adviser to Oxfam International, commented on the Copenhagen Draft:

> Like ants in a room full of elephants, poor countries are at risk of being squeezed out of climate talks in Copenhagen … This is only a draft but it highlights the risk that when the big countries come together the small ones get hurt.
>
> (*The Daily Star*, December 10, 2009)

Generally speaking, the articles demonstrated how the group led by the chief negotiator of the G77 made their voices heard and attributed responsibility for the impending failure of the summit to the leaders of the rich and powerful (or "other") countries. However, these influential sources also highlighted the problematic consequences of climate change for Bangladesh as well as the inadequacy of the actions to tackle these problems. For example, in 2015, the articles in the newspapers avoided framing the issue of climate change as a contestation between developed and developing countries, but rather highlighted the actions required to address climate change.

During the 2015 Paris Summit, bureaucratic sources were used in a similar but less confronting way. A case in point is: "G77 and LDC protest: Rich ask developing and underdeveloped to pay" (*Prothom Alo*, December 5, 2015). As Bangladesh's chief negotiator, Kazi Khaliquzzaman, stated: "Bangladesh has already spent 400 million dollars of its fund to tackle climate change. However, the fund for adaptation to climate change effect must be provided by the industrialised countries." This position was strongly supported by different activists, such as Shamshuddoha, a representative of the German-based Bread for the World, who commented that "this is a sort of trap by the industrialised countries to fail the Paris conference."

Journalists highlighted local climate change issues more prominently in 2015 than in 2009 in relation to the Bangladesh government's responsibility. They used information from bureaucrats to reinforce the country's increasing susceptibility to the effects of climate change. One example can be found in the coverage of coastal erosion due to sea level rise: "Ruined Tamarisk forest in Kuakata beach" (*Pothom Alo*, July 15, 2015). In this article, Mihir Kanti Dey, an official with the Coastal Forest Department, described how erosion due to rising sea levels was increasing rapidly and claimed that the beach could be completely inundated within the next ten years if immediate measures were not taken. The article also used two local residents as sources to describe the effects of the disappearance of trees as their homes and cultivable lands had become directly exposed to strong winds and large tidal waves. Similar concerns were echoed in the views of AKM Mostafa Zaman, a Dean at the Patuakhali Science and Technology University, who as an expert source emphasised that immediate attention was warranted to stop the erosion.

Overall, it may be said that, in the main, journalists cited activists and expert sources to verify the statements made by bureaucrats. However, the analysis clearly shows a shift in the journalists' focus. Unlike in 2009, powerful bureaucratic sources were used in 2015 not only to demonstrate the contestation between rich and poor nations, but also to cross-check the statements of climate victims (citizen sources). The process of verification included the Danish politicians' denial of the existence of a draft document and a counter to that from Di-Aping of the G77. This controversy indicated the use of a contestation frame between the developed and developing world. The following section further analyses the contestation between the two parties, using expert perspectives.

Expert sources: Demand for greater justice

The articles dominated by expert sources provided a diverse understanding of the purpose of using such sources in the news (Albaek 2011). By using expert sources who constitute an "efficient machinery of record" (Schudson 2006), the journalists mainly reinforced the extent of Bangladesh's vulnerability due to climate change and the inadequacy of the global climate framework. Although there were similar proportions of expert sources in the two periods, there was greater validation of experts' statements in 2015 than in 2009. Irrespective of the difference in validation, the most crucial subject matter that surfaced repeatedly during both study periods was Bangladesh's vulnerability to climate change and its lack of capacity to tackle the consequences of this global problem. A clear example of this theme can be found in the article: "Bangladesh worst affected by changing climate" (*The Daily Star*, December 8, 2009). This article was published following a report released by an NGO—Germanwatch—on the eve of the Copenhagen summit. It described Bangladesh's vulnerability to natural calamities, highlighting the intensity and frequency of the natural catastrophes that had occurred during the last 20 years and which had claimed 8,241 lives. This high frequency of natural disasters, which made Bangladesh one of the most vulnerable countries in the Global Climate Risk Index (GCRI), clearly suggested that "climate change" was responsible for the loss of thousands of lives in this country, a claim verified by a Bangladeshi-origin expert source. The article also mentioned the names of the two German authors of this report but it did not raise any questions or seek any comments from them about their findings on climate change. It did, however, question the process of measuring the GCRI using comments by Dr Saleemul Huq, an expert with the International Institute of Environment and Development, who said:

> It's really hard to make a climate risk index. Only the number of people killed in natural calamities and losses of properties were counted to make this report. But millions of people, who survived extreme weather events and who are suffering across the globe, were not taken into the account.
>
> (*The Daily Star*, December 8, 2009)

From this comment, it may be inferred that the article attempted to show the inadequacy of the GCRI by recognising the vulnerability of Bangladesh's climate victims (see also Roberts & Parks 2007). It may, however, also be interpreted as an indication of the different perspectives of experts from developed and developing countries on the risk index. While the German experts only counted the number of direct victims of calamities, their developing world counterparts also included the plight of those who survived. This may be why Dr Huq, the Bangladeshi-origin expert, argued that the extent of vulnerability encountered by this country was not comparable to that of other small nations because of the enormous scale of human suffering.

In sum, this article presented the German research report and cross-checked it, citing an expert source. This cross-checking was important because it enabled what lies beyond the immediate reality to be exposed. The local expert's perspective highlighted the study's inadequacy in failing to consider the plight of millions of disaster survivors. While this article demonstrated a contestation between international and local experts, the other articles also focused on the presence of various expert sources in climate change-related news. A case in point is: "Carbon cut pledge, legal string missing" (*The Daily Star*, December 20, 2009). This article emphasised the insignificant progress achieved at the Copenhagen Summit, which was reflected in the hurried signing of the deal at the very last minute. The deal did not include any major commitment to the reduction of carbon emissions from either developed (e.g., the US) or developing countries (e.g., China and India). The article cited one political and two expert sources. It sought comments from the previously cited expert, Dr Huq, who was somewhat critical of the outcome of the summit. He said: "It (Copenhagen) does not also specifically say about emissions curbing target of the developed countries. So, we can say Bangladesh did not get what we were expecting" (*The Daily Star*, December 20, 2009). However, politicians from Bangladesh, including the then Environment Minister Hasan Mahmud, felt reassured that the country had played a strong role in the summit, and that its "visibility" as one of the world's most vulnerable countries had been established. Although he termed this visibility as a "big achievement," the Bangladeshi minister also expressed his concern over the fact that the country had not received any commitment vis-à-vis the expected climate fund. These comments demonstrated that, while the politicians did not disagree with Dr Huq, they were inclined to see the outcome of the negotiations in a more positive light. This was similar to the inclinations of the politicians mentioned above in *The Daily Star* report. In another article, similar sentiments were expressed by expert sources: "Bangladesh successful in climate diplomacy, the battle is now for the compensation" (*Prothom Alo*, December 21, 2009). This article used three expert sources who were members of the Bangladesh delegation: two climate scientists and a member of the Intergovernmental Panel on Climate Change (IPCC). The IPCC member, Dr Atiq Rahman, commented: "Bangladesh has successfully utilised the scope of demonstrating its influence particularly in presenting scientific data about climate change which engendered an increasing sympathy towards us as one of the vulnerable nations in the world" (*Prothom Alo*, December 21, 2009). This recognition of vulnerability, together with the "uncertainty" surrounding the possibility of a climate fund, could be seen as assigning responsibility to "other countries." In particular, the achievement of "visibility" by Bangladesh demonstrated that the acknowledgment of susceptibility to climate change—and subsequent assistance to tackle it—actually depended on recognition by the developed countries.

The above analysis shows that the journalists positioned both experts and politicians in a synchronised, mutually reinforcing way, particularly in

Prothom Alo. This synchronised position raised further questions about the value of experts' viewpoints, which are supposed to be independent of the authorities (Ericson et al. 1989; Schudson 2006; Boyce 2006). *The Daily Star*'s article was a particular case in point: it used an international expert of Bangladeshi origin who was critical of the outcome of the summit. This expert's criticism raised serious questions about the views of the expert members in the official Bangladesh delegation, who expressed a very positive assessment of the summit outcome in accord with the official government position.

In these three articles, an important point to note is the difference of opinion among the experts. The use of one quite critical expert in two articles in *The Daily Star* and another expert's cautious observations regarding the conference outcomes in *Prothom Alo* could be regarded as indicative of subtle disagreement among the experts, the extent of which was not as explicit as in the articles dominated by bureaucratic sources. However, this disagreement was quite useful in that it exposed the differences in perspective between the experts from the developed and developing worlds, as well as between the international and local experts. However, the critical perspectives only applied to rich countries' commitment, which only strengthened the Bangladesh government's political position on climate change. Unlike the articles in 2009, the articles in 2015 emphasised expert and political sources in the action frame. This coverage highlighted the country's willingness to tackle the issue of climate change internationally (i.e., commitment to INDC). While politicians were the most frequently cited sources in 2009, in 2015, experts were important in the international negotiations on behalf of Bangladesh during the Paris meeting (COP21). Although the politicians were used as sources, they were also held responsible for not taking adequate action to mitigate climate change locally.

The experts' importance in 2015 was evident in an article titled "Bangladesh to reduce carbon emissions by 20 per cent" (*Prothom Alo*, November 28), in which Professor Ainun Nishat referred to the country's emissions reduction commitment in its INDC and claimed: "Despite the fact that Bangladesh has no role in global climate change, the country is committed to emissions reduction. Bangladesh should be an example for the rest of the world." By using this expert opinion, the article clearly invoked the action frame and highlighted the country's precariousness—despite its strong emissions reduction commitment—resulting from the inaction of rich countries. These quotes demonstrate that, despite a strong mitigation initiative from the affected countries, there was a lack of adequate action from the rich countries to tackle the imminent risks of climate change. This portrayal clearly reinforced the "north–south" divide (Chapman et al. 1997) and invoked the affected countries' demands for climate justice. While the affected countries' mitigation initiatives were established through the use of expert sources, there was an absence of international political or official sources in relation to rich countries' inaction. This gap can be explained by a lack of resources and lack of access to international sources in developing countries (Shanahan 2006).

Furthermore, the climate change coverage was often aligned with the political position of the government and purposefully displayed varied types of advocacy (Eide & Kunelius 2012). Here, the frequent use of expert sources allowed journalists to reaffirm the vulnerability of Bangladesh as a climate victim.

Activists and citizens: Emerging contesting forces

Activist sources

In 2015, environmental activists achieved a more dominant position in Bangladesh in the debate surrounding local climate change-related issues, such as the Rampal coal-fired power plant, local emissions reduction and climate aid. Activists were prominent in the Bangladeshi news media platforms "through which the majority of the public becomes aware of impending environmental threat" (Hutchins & Lester 2006, p. 434). These activists were predominantly verified by official and business sources. In 2009, the activists mostly operated locally and talked about both international and local climate issues. In 2015, however, they mainly collaborated with various international non-governmental organisations (e.g., BankTrack) to present more compelling evidence in support of their positions to save the environment.

A case in point here is "Questions over Rampal coal-fired power plant in Bangladesh" (*The Daily Star*, July 7, 2015), referring to BankTrack, a Netherlands-based coalition targeting private sector operations and investments and their effect on people and plants. The article stated: "In direct violation of the IFC (International Finance Corporation) requirement to comply with relevant laws of host countries, the Rampal project breaches Bangladesh's Environment Conservation Regulation 1997, the Environment Conservation Act 1995, and the Forest Act 1927." The BankTrack finding was cross-checked with the Managing Director of the Bangladesh–India Friendship Power Company Limited (BIFPCL) and Bangladesh's State Minister for Power, Nasrul Hamid, who rejected the findings, claiming that "People who are opposed to the project don't seem to know the details of it." Although the Managing Director claimed that the Environmental Impact Assessment (EIA) had been conducted, this was denied by internationally renowned environmentalist and water expert Professor Ainun Nishat. This claim and counter-claim can be explained by what Hansen (2011) refers to as the dialectical principle in the public sphere, whereby activist sources often provoke sharp counter-claims. Hansen also asserts that this principle is intensified when it meets the journalistic process of cross-checking that presents both sides of the debate but focuses on certain arguments.

At times, however, the activists' claims did not receive any cross-checking or stringent background scrutiny. For example, one journalist highlighted a "call for setting up international climate court" (*Prothom Alo*, December 2, 2015) during COP21. The call, made by local environmental activists during a rally in Noakhali district, was to implement legally binding emissions reduction targets for rich countries. However, the article neither verified the justification or practicality of the call via any other sources nor provided any contextual background for such an international court. Similarly, another article (*Prothom Alo*, October 16, 2015) highlighted Transparency International Bangladesh's (TIB) Executive Director Iftekharuzzaman's demand for compensation from rich countries so as to enable Bangladesh to implement adaptation projects that had been experiencing funding constraints. The article also alluded to the debate on climate loan vs climate fund, in which Iftekharuzzaman argued strongly for a climate fund in the form of compensation, since Bangladesh was not responsible for emissions but, rather, a victim of their consequences. He further stated: "It is immoral to use climate fund as a profit generating investment opportunity." As in the previous article, this journalist did not cross-check with any relevant authorities, leaving aside a significant concern for climate fund.

Although these articles did not provide any further information about climate aid to Bangladesh, the journalists did mention "mistakes" and "waste of money" in relation to some climate funds, such as the Bangladesh Climate Change Trust Fund, Bangladesh Climate Resilience Fund, Palli Karma Shahayak Foundation, and Pilot Project on Climate Resilience (Huq 2015). These mentions did not, however, amount to any critical engagement with the issue of a climate fund during the study period. One possible explanation of this lacuna is that the journalists were keen to establish the country's vulnerability and ensure appropriate funding for climate action through the global climate policy framework, thus protecting the government's interest, instead of exposing any operational weaknesses of the existing climate funds in Bangladesh in the wake of the global summits.

This explanation, however, does not seem to be applicable to the debate concerning the Rampal Power Plant, which resurfaced during the Paris summit. A case in point here is the headline: "Slogan in Paris: Save the Sundarbans" (*Prothom Alo*, December 8, 2015). Referring to one of the activists, the article focused on the slogan "Save the Sundarbans, Save Tigers, Stop Rampal." The activist's assertions were cross-checked with the Environment Secretary of the Bangladesh government, Kamaludddin Ahmed, who added that the proposed site of the power plant was approximately 14 kilometres away from the outer boundary of the Sundarbans and 65 kilometres from this world heritage site. The quote from this official was a denial of the environmental threat posed by the proposed plant. To scrutinise the official sources, the journalist seemed to have privileged the activists' viewpoints by portraying their demand explicitly in the headline. The cross-checking with the official showed a clear contestation between the two sources, and the article demonstrated that the journalist stood by environmental causes without being tendentious (Manning 2000). While activist sources were prominent in these newspapers, citizen sources or ordinary people appeared less frequently. The following section provides an overview of the paucity of citizen sources which, somewhat surprisingly, were used in a frame similar to that of political sources.

Citizen sources

Despite journalists' professional preference for authoritative sources (Ericson et al. 1989), citizen sources add significant value to climate change news. Yet, in numerical terms, ordinary people were not as visible as other sources in the two newspapers' coverage of climate change during both study periods. Nonetheless, the presence of this particular type of source reveals the significance of some local issues which indicated that climate change was part of an on-going environmental struggle in Bangladesh. In 2015, citizens were used as direct victims of erosion and flooding due to a rise in sea level. In 2009, they were used to highlight how some traditional occupations, such as fishing and agriculture farming, were struggling to adapt to climate change. For example: "Farmers never say over in climate change war: Try out cope-up measures as cropping seasons shift" (*The Daily Star,* December 5, 2009). This front-page article presented the views of five farmers and a seed trader in relation to the onset of climate change in Bangladesh, a phenomenon manifested in the shifting monsoon season and rapid encroachment of salinity on arable lands in the country's southern regions. This process of salinity, which had spread to cover 1.2 million hectares of land by 2009 from 0.83 million hectares in 1990, demonstrated the extent to which erratic climatic patterns were impacting on Bangladesh's farming communities. Septuagenarian farmer Abu Bakar Siddiqui of Louhoni village reported that, compared to previous years, winter now came late and was shorter in duration and that the difference between day- and night-time temperatures had been widening. The elderly farmer's views were supported by a local seed trader, Muhammad Shah Alam, who frequently received complaints from farmers that they had applied similar amounts of pesticides as in previous years but were "not getting relief from pest attacks."

Dr S. K. Ghulam Hussain, a Director of the Bangladesh Agricultural Research Council (BARC), who had been monitoring the possible impact of climate change on the country's farmlands, confirmed Abu Bakar's concerns and stated: "We're in a virtual climate change laboratory, which is a natural advantage for us." The term "climate change laboratory" suggested that the changing nature of the climate made the people more open to innovation, and more able to adapt to the consequences of increasing temperatures and shifting seasons. This malleability was also reflected in the farmers' switch from the rice variety BR-28 to BR-29, which has the capacity to cope with temperatures that exceed 30 degrees Celsius, a fact confirmed by rice-breeding scientist Professor Zeba Seraj of the University of Dhaka.

While the farming communities were seemingly fighting the adversities of climate change and winning, the scenario was gloomy for the fishermen in the country's south-western coastal regions, whose livelihoods were becoming seriously endangered. The article, "Changing climate, disappearing trade" (*Prothom Alo,* December 7, 2009) cited 56-year-old fisherman Abdul Jalil, who had been fishing in the Bay of Bengal for the last 35 years. He said that he had never seen so many natural disasters in his lifetime. During the past year, the conditions were so bad that he was unable to fish, not even for a month. Another fisherman, Syed Nur, said that fishing communities in the coastal zone were being severely affected by natural disasters. When inclement weather prevented the fishing trawlers from going to sea, there was no income for the fishermen. If a trawler sank and the fishermen drowned, there was no compensation for their families. With this in mind, many fishermen had opted to quit fishing and had commenced day labouring to support their families, Nur added. The fishermen's frustration over the changing climate and their struggle against natural calamities were explained by two expert sources: Md Rahseduzzaman of the Bureau of Meteorology and Professor A. Q. M. Mahbub of the Disaster Research Centre at the University of Dhaka. Both experts attributed the plight of the fishermen to the rising water temperature in the Bay of Bengal, which caused frequent periods of low pressure and cyclones in the coastal areas. The increasing frequency and intensity of pressure was primarily caused by climate change over the last few years.

The analysis of the articles in 2015 revealed that the citizen sources were verified predominantly by bureaucratic rather than expert sources. These bureaucratic sources were used as validation in the coverage of the effects of climate change, such as coastal flooding and erosion. Although the authors of these articles described the suffering of local people, they did not raise any questions about widely-discussed issues such as infrastructural challenges to Bangladesh's ability to tackle climate change (Dastagir 2015; Islam & Islam 2016; Rahman 2018). In 2009, the articles established the effects of a changing climate by exposing two traditional yet still significant occupations in this riverine country, namely, agriculture and fishing. Through reporting and verification, the articles revealed that while the farming community was adapting, the fishermen were being displaced. Verifying citizen sources who were experiencing climate change by using "knowledgeable" expert sources, the newspapers sought to further highlight the vulnerability of the country. The position on climate change was legitimised through the example of the adaptation of farming and the displacement of fishing.

This focus on adaptation continued in the articles from 2015, but now official sources were at the forefront of the cross-checking process. By prioritising a top-tier source (Martin 1997) in the validation process, the journalists focused on the efficacy of the adaptation programmes in this country. The sparse use of citizen sources, who usually have less access to the media (Lewis et al. 2008), was in contrast to the profound impact of their excerpts in demonstrating the effects of climate change in Bangladesh. This contrast demonstrated how the journalists were able to direct their influence towards bringing forward some crucial issues. In this case, the journalists sought to frame climate change as a problem in order to demonstrate how Bangladesh was trying its best to adapt.

Business sources: Managing emissions

While the number of articles dominated by business sources was small, the tone of these articles was positive towards various climate change issues, predominantly those related to mitigation. Business voices were quoted in an action frame that emphasised how these interests were taking the initiative in the global process of emissions reduction. However, as stated above, the share of articles dominated by business sources was very low in both study periods, although there was a slight increase in 2015.

Despite this low quantitative presence, the articles were significant in as much as they dealt with the contentious mitigation issue in the carbon trading scheme, which allowed countries or companies to buy or sell "carbon credits." The scheme was introduced to reduce carbon dioxide in the atmosphere. While the media representation of carbon trading was contentious in Australia (see previous chapter), it received very positive coverage in Bangladesh, which is evident in the following analysis of two articles: "Brick kilns going green: Operators eye benefit from carbon trading" (*The Daily Star*, December 14, 2009) and "COP15 Bangladesh is selling 'economic carbon'" (*Prothom Alo,* December 13, 2009). While *The Daily Star* article was based on a visit to brickfields near Dhaka by the correspondent, the *Prothom Alo* article was about the COP15 conference in Copenhagen and highlighted the fact that the delegates' air travel would release 40,000 tonnes of carbon dioxide into the atmosphere. Based on this figure, the Danish government had planned to issue financial instruments/certificates worth $15.20 per unit of carbon pollution for this conference. *The Daily Star* article, which referred to the Dhaka-based brick manufacturing company, asserted that the company was eligible to receive a certificate because the clean technology that it used in the brickfields released far fewer emissions into the atmosphere. This particular brick business was funded by various industry finance companies. Motiul Islam, the chairman of one of the companies, hoped that the business would earn "significant foreign currencies" through carbon trading that supported this new technology and he had recommended it to the brickfield owners in Dhaka. The article considered this company to be "environmentally friendly."

The above excerpt demonstrated that the brickfields—one of the highest polluting industries in Bangladesh—could cut their carbon emissions significantly by using the new technology. However, none of the assertions was verified by any expert sources or by the relevant Danish Authorities or the World Bank. Given this absence of verification, the articles could not shed any convincing light on the prospects for "carbon trading" in Bangladesh. This is particularly important in the context of a mounting global controversy around the efficacy of carbon trading, a topic that was discussed in detail in the previous chapter on news sources in Australia. Notwithstanding this absence, the articles quoted the business sources who clearly spelled out the company's enthusiasm for business opportunities that could help Bangladesh to reduce its carbon footprint both locally and globally.

Similarly, in 2015, business sources were used in an article related to the World Bank's collaborative project with the Asian Development Bank and the German Development Bank to support two local organisations to implement the Carbon Credit Program of the United Nations Framework Convention on Climate Change (UNFCC). The implementing organisations were Grameen Shakti of Noble Laurate Professor Yunus, and the state-owned Infrastructure Development Company Limited (IDCOL). In reference to an IDCOL research study, its Director (Investment) Nazmul Haque said:

> a family needs eight litres of kerosene a month for lightening up a house in a rural area ... As we have already installed four million solar home systems, this programme is saving around 32 million litres of kerosene every month ... The IDCOL programme saves 2.7 lakh tonnes of kerosene annually, cutting about 4.88 lakh tonnes of carbon emissions in the process every year.

The director also offered assurances that the World Bank would be the main procurer of their certified emissions reduction units. The article checked with two solar home system owners, who were happy to replace their kerosene lamps with the solar system lighting. It did not, however, incorporate any statement from the World Bank regarding the procurement process. Bangladesh currently generates 2.5 per cent of its electricity from renewable sources, and has a target of increasing renewable electricity production to 10 per cent by 2020.

However, some statements from the business sources were not verified, an omission that raises some questions. including whether the journalists deliberately presented a positive image of both the company and Bangladesh, and whether the non-verification of the business sources reflects the reluctance of this type of source to engage with the usual process of scrutiny of mainstream journalism (Galbraith 2004, as cited in Doyle 2006; see also Davis 2002). By assigning responsibility to "other nations" without incorporating the views of the implementing authorities regarding carbon trading, the journalists sought to demonstrate that, although Bangladesh was not wholly responsible for carbon emissions, it could assist the world to reduce emissions into the atmosphere. In this way, it adopted a positive albeit superficial stance by presenting action to tackle climate change as a "new business opportunity" (Painter 2016).

Conclusion

The use of sources and excerpts allows journalists to call on a range of voices, such as politicians, citizens, experts and bureaucrats, to frame stories based on newspapers' precise arguments (Pan & Kosicki 1993; Schneider 2011). The analysis of various principal sources demonstrated that the framing of climate change had shifted from climate change as a problem and a matter of contestation between rich and poor nations, to an action frame in which all parties were united in seeking to tackle the effects of climate change. There was also more cross-checking of sources in 2015 than in 2009, although the number of sources had decreased in the latter period. This decrease in the number of sources can be explained by the concept of news values, which underpins professional journalistic culture. While the Bangladesh delegation to the widely covered Copenhagen summit (Painter 2016) was led by the Prime Minister, the country's highest political authority, accompanied by a diverse team of experts, officials and politicians, its Paris delegation was led by the Environment Minister, a relatively low-profile political figure. This change in delegation leadership essentially influenced the news value of the coverage for the two Bangladeshi newspapers, resulting in a lower level of coverage with enhanced cross-checking. This enhanced cross-checking can be explained by two factors. First, in the cacophony of a "web dominated information marketplace" (Shapiro et al. 2013, p. 657), the journalists might have wanted to present a more professional, standalone product. Second, the absence of high-level influential sources might have provided journalists with more room to cross-check the sources. As mentioned before, although it is not possible to cross-check all daily events, the process is inherent in the "hypothesis-testing feature" (Pan & Kosicki 1993, p. 61) of news, which makes it imperative to provide statements from the other side to illustrate the point, thus consolidating the frame.

One of the significant findings that warrants further attention the presence of non-dominant sources, namely, activists and citizens in this uneven playing field (Hansen 2011). These sources were also present as principal sources in 2009, but were more prominent in 2015 in relation to both local and global climate change issues. The use of these sources demonstrated the dynamic relations between journalists and activists that are reflected in the trade of information between them (Broersma et al. 2013). This trade took the form of information provided by activist sources and the preferential treatment offered by journalists via their positioning and critical verification of the activists' statements. Together, verification and selection—which are integral aspects of journalistic practice—invoke the "dialectical principle of debate in the public sphere" (Hansen 2011, p. 12), engendering claims by the activists and counter-claims by the political and official sources. Here, the number of activists was of no import; what was crucial was the journalistic process of selection and salience that revealed the sources' capacity to challenge the authorities, and thus exposed some critical environmental issues in the public domain.

The prominence of political sources in 2009 demonstrates how the public debate alluded to the issue of climate justice. This issue may be viewed as emanating from the country's vulnerability. However, in 2015, the presence of less dominant sources (i.e., citizens and activists) exposed a crucial matter besides climate vulnerability (e.g., rising sea level) that was reinforced by the citizen sources. The activist sources finger-pointed at potential future catastrophes (e.g., Rampal plant). While the reinforcement of vulnerability strengthened the political position of Bangladesh regarding climate justice, the lack of attention to dominant sources (political, bureaucratic) in the representation of local potential climatic hazards can be explained through what Beck (2016, p. 98) considered as "problematic of invisibility," which is innately connected to the "problematic of power." In this power relation, the journalist potentially has the means to make the invisible threat of environmental harm from the coal-fired power plant more visible by drawing our attention to those imperceptible risks.

References

Albaek, E. 2011, 'The interaction between experts and journalists in news journalism', *Journalism*, vol. 12, no. 3, pp. 335–348.

Anderson, A. G. 2014, *Media, environment and network society*, Palgrave Macmillan, Basingstoke.

Anderson, B. 1991, *Imagined communities*, 2nd edn, Verso, London.

Beck, U. 1992, *Risk society: towards a new modernity*, translated by M. Ritter, Sage, London.

Beck, U. 2010, 'Climate for change or how to create a green modernity', *Theory Culture & Society*, vol. 27, no. 2–3, pp. 254–266.

Beck, U. 2016, *The metamorphosis of the world*, Polity, Cambridge.

Boyce, T. 2006, 'Journalism and expertise', *Journalism Studies*, vol. 7, no. 6, pp. 889–906.

Broersma, M., den Herder, B. & Schohaus, B. 2013, 'A question of power: The changing dynamics between journalists and sources', *Journalism Practice*, vol. 7, no. 4, pp. 388–395.

Chapman, G., Kumar, K., Fraser, C. & Gaber, I. 1997, *Environmentalism and the mass media, the north-south divide*, Routledge, London, New York.

Cottle, S. 1998, 'Ulrich Beck, "risk society" and the media: A catastrophic view?', *European Journal of Communication*, vol. 13, no. 1, pp. 5–32.

Dastagir, M. R. 2015, 'Modelling recent climate change induced extreme events in Bangladesh: a review', *Weather and Climate Extremes*, vol. 7, pp. 49–60.

Davis, A. 2002, *Public relations and democracy*, Manchester University Press, Manchester.

Doyle, G. 2006, 'Financial news journalism: a post-Enron analysis of approaches towards economic and financial news product in the UK', *Journalism*, vol. 7, no. 4, pp. 433–452.

Eide, E. & Kunelius, R. (eds) 2012, *Media meets climate: The global challenge for journalism*, Nordicom, Göteborg, Sweden.

Ericson, R. V., Baranek, P. M. & Chan, J. B. L. 1989, *Negotiating control: a study of news sources*, Open University Press, Milton Keynes.

Galbraith, J. K., 2004, *The Economics of innocent fraud: Truth for our time*, Allen & Unwin, London.

Hansen, A. 2011, 'Communication, media and environment: towards reconnecting research on the production, content and social implications of environmental communication', *International Communication Gazette*, vol. 73, no. 1–2, pp. 7–25.

Hutchins, B. & Lester, L. 2006, 'Environmental protest and tap-dancing with the media in the information age', *Media, Culture & Society*, vol. 28, no. 3, pp. 433–451.

Huq, S. 2015, 'The inside story of the Paris agreement', *The Daily Star*, December 15, accessed December 21, 2018, available: https://www.thedailystar.net/op-ed/the-insi de-story-the-paris-agreement-187159.

Islam, M. S. & Islam, M. N. 2016, 'Environmentalism of the poor: The Tipaimukh dam, ecological disasters and environmental resistance beyond borders', *Bandung: Journal of the Global South*, vol. 3, no. 1, pp. 1–16.

Kovach, B. & Rosenstiel, T. 2007, *The elements of journalism: What news people should know and public should expect*, Three Rivers Press, New York.

Lewis, J., Williams, A. & Franklin, B. 2008, 'A compromised fourth estate? UK news journalism, public relations and news sources', *Journalism Studies*, vol. 9, no. 1, pp. 1–20.

Machill, M. & Beiler, M. 2009, 'The importance of the Internet for journalistic research: A multi-method study of the research performed by journalists working for daily newspapers, radio, television and online', *Journalism Studies*, vol. 10, no. 2, pp. 178–203.

Manning, P. 2000, *The sources of social power*, Sage, London.

Martin, M. 1997, *Communication and mass media: Culture, domination, and opposition*, Prentice Hall, Scarborough, Ontario, Canada.

Nassanga, G., Eide, E., Hahn, O., Rhaman, M., Sarwono, B. 2017, 'Climate change and development journalism in the global south', in E. Eide, R. Kunelius & K. Kumpu (eds) *Global climate, local journalism: A transnational study of how media makes sense of climate summits*, Projekt Verlag, Freiberg, Germany, pp. 213–233.

Neuman, R. W., Just, M. R. & Crigler, A. N. 1992, *Common knowledge: News and the construction of political meanings*, University of Chicago Press, Chicago, IL.

Painter, J. 2016, 'Disaster, uncertainty, opportunity or risk? Key messages from the television coverage of the IPCC's 2013/2014 reports,' *METODE Science Studies Journal*, no. 6, pp. 81–87.

Pan, Z. & Kosicki, G. 1993, 'Framing analysis: An approach to news discourse', *Political Communication*, vol. 10, no. 1, pp. 55–75.

Park, R. E. 1940, 'News as a form of knowledge: A chapter in the sociology of knowledge', *American Journal of Sociology*, vol. 45, no. 5, pp. 669–686.

Roberts, J. T. & Parks, B. 2007, *A climate of injustice: Global inequality, north-south politics, and climate policy*, MIT Press, Boston, MA.

Rahman, M. A. 2018, 'Governance matters: Climate change, corruption, and livelihoods in Bangladesh', *Climatic Change*, vol. 147, no. 1, pp. 313–326.

Schneider, J. 2011, 'Sourcing homelessness: How journalists use sources to frame homelessness', *Journalism*, vol. 13, no. 1, pp: 71–86.

Schudson, M. 2006, 'The trouble with experts – and why democracies need them', *Theory and Society*, vol. 35, no. 5–6, pp. 491–506.

Semetko, A. H. & Valkenburg, M. P. 2000, 'Framing European politics: a content analysis of press and television news', *Journal of Communication*, vol. 50, no. 2, pp. 93–109.

Shanahan, M. 2006, 'Science journalism: Fighting a reporting battle', *Nature*, no. 443, pp. 392–393.

Shapiro, I., Brin, C., Bédard-Brûlé, I. & Mychajlowycz, K. 2013, 'Verification as a strategic ritual: How journalists retrospectively describe processes for ensuring accuracy', *Journalism Practice*, vol. 7, no. 6, pp. 657–673.

Tuchman, G. 1978, *Making news: A study in the construction of reality*, Free Press, New York.

van Dijk, T. 1991, *Racism and the press*, Routledge, London.

7 Australia and Bangladesh: Conclusions

The preceding chapters explored how journalists deployed two different forms of deliberations—a fact-based news style and an opinion-based commentary style—in the coverage of climate change. These two styles of deliberative practices enabled journalists to bring various actors and processes together to construct the issues surrounding climate change in relation to their respective national interests. National interests are important considerations because some problems pertaining to climate change are resolvable only through local political interventions. For example, clean energy initiatives in the global North (i.e., Australia) and continued demand for climate justice from the global South (Bangladesh) emanate, to some extent, from lack of good governance (Bhuiyan 2015; Sovacool 2017) in the local context alongside inherent structural inequalities in the global context (Roberts & Parks 2007). The overarching demand for environmental justice from both national and international parties in climate negotiations potentially resonates with the notion of burden-sharing justice underpinned by "the principle that those who have caused the problem should bear the burden" (Caney 2014, p. 125). The evidence presented in the previous chapters demonstrated the importance of examining the factors that have enabled the exploration of similarities and differences in the selected countries' news coverage, and the extent to which diversity and uniformity of journalistic practices (Golding 1977; Esser & Hanitzsch 2012) were identified in these two diverse countries. Since the discussion is based on qualitative analysis, it is worth mentioning that the comparison between the two countries is reflective of the particular cases rather than representative of all potential cases. Analysis of similarities and difference (discussed in Chapters 2 and 4) contributed to complex climate change discussions (Hulme 2009) in the intersection of global and local perspectives.

In Australia, the newspaper coverage demonstrated that the local and global issues of climate change were strongly linked. This link was observed at both the policy and scientific levels, and was palpable in the coverage of the thorough scientific investigations conducted by the CSIRO and the Bureau of Meteorology and of the policy issues around clean energy. In Bangladesh, however, the extent to which these local and global climate change issues were interrelated differed. Fewer Bangladeshi newspaper articles addressed environmental issues and, in these articles, global and local issues were seldom related. This difference in the pattern of relationship has its genesis in the diverse nature of environmental problems faced by the two countries. Australia is an arid continent, while Bangladesh is a low-lying "land of rivers." Apart from these natural characteristics, there are also important differences in their respective political and media contexts. The difference in the news coverage corroborates James Carey's (2007, p. 5) contention that the practice of journalism is "a craft of place" and shows how news media operate in the "light of local knowledge."

In the articles that appeared in the two newspapers from Bangladesh, the pressing issue was severe exposure to climatic vulnerability and the need to deescalate the impact of exposure through "transformational adaptation" (Huq 2019) measures. Here, the broad issue of climate change (e.g., increasing emissions) mattered less compared to the micro-level climatic problems of increased salinity and the impact of rising sea levels on remote communities. The focus on potential areas of ecological degradation indicated that the inherent characteristics of environmental coverage were underpinned by the "alarmed discovery" of the problems and the news media's willingness to present solutions to said problems (Downs 1972; Hansen 1991). The identification of local news topics, such as Tipaimukh and Rampal, offered useful pegs for journalists to highlight these issues in the public domain, but the nature of the inter-relation between the global and local climate change is only implicit. Drawing on Beck's (2009, p. 11) notion of risk, it may be said that the representation displays a distinction between anticipated catastrophe (e.g., the prospect of a dam upstream or a coal-fired power plant near a forest) and actual catastrophe (rising sea level). Bangladeshi journalists focused on both the anticipated and actual risks to alert global and local authorities to the need for preventive measures. However, the pattern of attention in Bangladeshi newspapers was similar to that in Australia in terms of sustained attention to the national interest, although this does not mean that the editorial positions of all four newspapers were the same. Rather, they differed in as much as Australia saw itself as a developed country taking the initiative to reduce its emissions through clean energy investment, while Bangladesh as a developing nation saw itself at the receiving end of—and wronged by—global emissions. Within these broad positions, the journalists maintained their particular focus on the respective countries' climatic problems. In Bangladesh, they sought climate justice in cases of both local development projects and unrelenting global climatic hazards. These writers sought justice from the greenhouse-gas-emitting Western countries, a position more prevalent in the articles pertaining to the Copenhagen summit than in those about the Paris conference. Similarly, in their articles concerning Tipaimukh and Rampal, the journalists sought justice from relevant national governments to act on salvaging the county from potential anthropogenic climatic disasters. In Australia, the news media generally focused on the leadership contestation over climate policy. Here, the differences between the two newspapers were more marked, although, in broad terms, their concerns regarding climate policy were somewhat similar. This similarity was evident in the proposed "market solution" (e.g., carbon permit) to address broad issues of climate change in Australia. However, the orientation towards these matters differed between the two newspapers, which can be considered as two sides of the debate. One side (articles in *The Sydney Morning Herald)* evinced a "concerned" position because the government was not doing enough to tackle climate change. The other side (articles in *The Australian)* demonstrated an "unconvinced" position vis-à-vis market solution policies (i.e., the Emissions Trading Scheme, the Renewable Energy Target, etc.) because of their potential conflict with Australia's economic interests. Here, the newspaper is essentially invoking the "language of risk" (Painter 2013, p. vii), paving the way for public debate to shift and policy action to be stalled until there is conclusive evidence showing these climate policies would not hurt national economy.

The "unconvinced" position of *The Australian* drew some critical attention from the fields of political science (Manne 2011) and journalism studies (Chubb & Bacon 2010; Bacon 2011). Robert Manne's investigation clearly revealed how the opinionated columns in *The Australian,* which were predominantly written by "sceptical non-scientists," denigrated views in favour of taking radical action to tackle climate change (McGaurr & Lester 2013). During the Paris summit, a sort of consensus on the point of market solution (i.e., business-driven transformation) is evident in both the Australian newspapers, although these articles overlooked Australia's responsibilities and obligations (except few sympathetic articles on climate aid in the *Sydney Morning Herald*) beyond national interest, despite the country being "one of the very highest of all countries in terms of emissions per person" (Climate Council of Australia 2015, p. 23). Based on the evidence, it can be asserted that the Australian newspapers were influenced by what critics have called a "spatio-temporal" framework (Harvey 1996, p. 264) that "represents a particular set of values about both environment and society which privileges certain places, economic sectors, and people." This influence is exemplified in the prioritising of the resources sector or concerns about potential job losses found in the Australian content. Similarly, the influence of the framework can be identified in the Bangladeshi content in the form of a debate concerning environment versus development (i.e., Rampal) or hope for selling carbon certificates by local companies.

The findings are consistent with those of some comparative studies of environmental issues in the US, UK and France (Brossard et al. 2004; Boykoff 2007). These studies argued that the practice of journalism in these countries differed significantly, given that the news content was "domesticated", despite the fact that climate change has no borders. While these studies examined environmental issues and the professional norm of objectivity (see Chapters 3 and 5), the current study has extended the focus to the use of sources to understand their role in the framing of "policy debates" surrounding climate change.

Climate change and news sources

Comparison of news coverage related to global climate change policy reveals a broad similarity between Australia and Bangladesh in the prominence given to political sources, mainly due to the transnational nature of the issues (Tuchman 1978; Hall et al. 1978; Ericson et al. 1989; Eide et al. 2009; Boykoff 2011). However, most visible political sources in Australia were also less verified. This lack of verification may have stemmed from daily journalistic constraints as well as the reliance on "hierarchies of credibility" (Becker 1967; Carlson 2017), paving the way for amplification of certain powerful voices over others. In the continuum of public knowledge, the defining factors of credibility are under-pinned by the contestation between two opposing forces in Australia. This contestation is only assumed in the context of the relative absence or invisibility of particular social agents, e.g., activist sources, despite the fact that Australian journalists make "man-made" and "incalculable" risks explicit predominantly through activist sources (e.g., in the news articles pertaining to the Great Bar-rier Reef). In Bangladesh, too, the climate coverage was dominated by political sources and the contestation between politicians and other sources was less explicit. The public knowledge about climate justice was indeed defined by powerful political forces, but the presence of activists or civil society forces was also significant. Recently, the activist sources had gained more access and sought to intervene in the news media's process of amplification.

Beck's notion of climate risk helps to make sense of the corresponding nature of the representation of climate change. The Australian newspapers used sources to portray the anticipation of future catastrophes in order to prevent them (e.g., renewable energy resolutions). Similarly, although Bangladesh was a peripheral force in the climate change negotiations (Eide & Ytterstad 2011), journalists in this country also pointed a finger at other countries and held them responsible for the failure of the talks in Copenhagen. However, unlike in Australia, the newspapers in Bangladesh cited national political leaders in an action frame to promote the issue of environmental justice. The country's leaders campaigned for this view to influence the discussion via a recommendation regarding "cli-mate refugees," according to which climatically displaced people should be per-mitted to seek refuge in developed countries (see Chapters 5 and 6). In effect, both newspapers in Bangladesh defined the terms of the international debate on climate talks by citing political sources.

The Australian newspapers, which perceived themselves as representative of one of the world's developed countries, sought to influence the signing of a global climate deal. The Bangladeshi newspapers, which represent a small developing nation, demanded justice from the rich countries that they held responsible for emissions across the globe, and for victimising innocent countries like Bangladesh with their relentless emissions (Giddens 2011). In both countries, political sources were found to be verified less frequently than other principal sources. In Bangladesh, however, verification was carried out to a far less extent. All four newspapers attempted to maintain a positive coverage for their respective countries (predominantly in the case of global climate negotiations) by providing uncritical space to their politicians, indicating the strength of the relationship between journalism and the particular nation states in question (Anderson 1997; Eide & Ytterstad 2011). Curran's (2002) notion of a "restricted prism" implies the degree to which news producers are inclined to focus on issues through the lens of the nation state and locality.

The relative lack of verification of political sources in Bangladesh gave rise to several questions. First, because Bangladeshi newspapers had limited access to influential international sources, for example at the Copenhagen summit (Shanahan 2006, 2011), they reported mainly on specific *events*, rendering the description of what happened as adequate. Second, Bangladeshi newspapers may have concluded that the issue of utmost significance was to inform its audience about Bangladesh's strong role in establishing the footprint of climate justice. In the process of informing the audience, political sources went unchecked, perhaps because of diverse operational and professional factors that can be seen as impediments to the development of Beck's cosmopolitan vision (Wimmer & Quandt 2006; Cottle 2009; Laksa 2014).

Experts and bureaucrats were the second most frequently cited sources in both Australia and Bangladesh, a finding consistent with a single-nation study of the Bali summit undertaken by Eide and Ytterstad (2011). Statements made by expert sources were verified more frequently in articles in *The Australian* than in *The Sydney Morning Herald*. Scrutiny of the experts' positions through the journalistic norm of verification was consistent with *The Australian*'s sceptical position on climate change (McKnight 2010; Nash et al. 2009). Somewhat ironically, this scrutiny revealed how "journalism of verification" (Kovach & Rosenstiel 2007) was used to encourage doubt about climate change issues (see also Boyce 2006; Boykoff 2004, 2011).

In Bangladesh, articles dominated by experts were verified a little more frequently in the English daily than in the Bangla newspaper. Unlike in Australia, both newspapers in Bangladesh strongly endorsed climate change as confirmed by numerous scientific investigations (Raupach et al. 2008). The articles argued that the extent of Bangladesh's vulnerability had not been represented adequately, thereby exposing the differing viewpoints between the experts from developed and developing countries. The reasons for highlighting these differences between the experts of two different regions may be explained in two ways: (1) the experts attempted to establish Bangladesh's political position regarding the extreme vulnerability of this low-lying delta land; and (2) they endeavoured to challenge "Western eco-imperialism", whereby the developed countries tended to enforce their views on developing countries without taking into full consideration the enormity and impact of the latter's different problems. One good example of this neglect was the impact of natural calamities on Bangladesh's environment and society (Driessen, 2002; Hannigan 2006; Giddens, 2011).

To generalise, in Australia, the experts were cited to show the impact of and various solutions to address the challenges emerging from the proposed climate policies. One of the significant points to note between the two periods was that while in 2009, the experts were used to question scientific evidence related to climate change, in 2015, they were used to question the Australian government's proposed climate policy and the lack of initiative to tackle the challenge (e.g., the security challenge emanating from climatic disasters). Conversely, the experts who were cited in the Bangladeshi newspapers unanimously endorsed the government's political position regarding climate change. As well, they expressed unequivocal support for the scientific findings on climate change. Interestingly, these sources provided ample support to establish the degree of the country's vulnerability, while in 2015, they also raised concerns about the government's emissions-intensive energy policy (e.g., the decision to construct the Rampal power plant). Between 2009 and 2015, both Australian and Bangladeshi newspapers witnessed an increase in the number of expert sources, which is similar to the trend in Danish newspapers (Albaek at al. 2003). The "relations of definition" that informed Bangladeshi newspapers' portrayal of climate risks and consequences relied on the dominance of expert sources, without any significant contestation of their perspectives. As a result, journalists and experts used each other in a mutually beneficial manner: journalists tried to frame the media debate by citing and interpreting experts to fit their environmental agenda; and experts utilised the relatively easy access granted to them to build their respective climate perspectives in the broader social debate. In contrast, both the Australian newspapers challenged experts' legitimacy on the pretext of protecting national economic interests. One may explain these scenarios by claiming that, in Bangladesh, the experts were politically close to the government positions in regards to global climate policy, but they were willing to challenge the authorities in terms of local issues (i.e., Rampal). In Australia, the newspapers did not take for granted the expert opinions on both local and global climate change issues. Rather, the articles contested the experts' legitimacy by invoking the journalistic professional norm of balance. The findings indicate that none of the articles examined in this study, except a few editorial opinions from Bangladesh, challenged the experts' knowledge in exposing the underlying antagonism between "those afflicted by the risks and those who profit from them" (Beck 1992, p. 46).

Significantly, in regard to less frequently cited sources, the Australian newspapers used activists to legitimise their respective positions concerning climate change, whereas Bangladeshi publications used citizens. In Australia, however, both newspapers legitimised their different positions on climate change by citing their sources expediently. In Bangladesh, the two publications used citizen sources to convey the impact of climate change on the country's two traditional sectors, farming and fishing. The use of citizen sources reinforced the experts' predictions about climate change in two different ways (see Chapter 2). While the articles in the English newspaper portrayed Bangladesh as a "climate laboratory" wherein farmers were adapting well, the Bangla daily articles emphasised how the increasing rate of natural calamities was causing some traditional traders, such as fishermen, to lose their livelihoods. The use of the citizen source in the action framing suggests that the news media were attempting to reify the process of adaptation; that is, to show that Bangladeshis were not just sitting on their hands waiting for aid from the developed countries, but rather quite actively mitigating climate change (Giddens 2011).

Implications

In the preceding sections, I have compared the use of sources in the coverage of climate change to show a diverse picture of the contrasting journalistic practices in the two countries. In Chapter 4, the context was set by laying out the prominence of issues in two dissimilar environmental contexts. Chapters 5 and 6 provided a detailed analysis of the journalistic strategies of the use of sources in relation to the framing of the crucial issues of public interest. These strategies, which were examined through verification of source statements, enabled an understanding of the significance of such a comparison of journalistic practices at various levels: Australia as an advanced industrialised country and Bangladesh as an economically less resourced nation.

Newspapers' implicit political positions

The comparison of climate change coverage has demonstrated that politicians were used as dominant sources in both countries' content. It also showed the pattern of source dominance in stark contrast to that of the framing of sources. Although the dominant sources were similar in both countries, they were used in totally different framings. While the politicians in Australia were frequently presented in the contestation between political forces, in Bangladesh the use of political forces was predominantly underpinned by shared interest (Snow & Benford 2000) and calls for action to invoke climate justice. This contrast suggests that the framing of sources is a "political strategic tool" (Entman et al. 2009) employed by publications to shape public dialogue surrounding the climate change issue and to give meaning to the diverse phenomena that emerge from the different political, media and environmental contexts explored in Chapter 2.

The difference in the framing of political sources may be due to the influence of a complex media system on journalists (see Chapter 2). News is always a joint production of journalists and sources. While journalists belong to particular media companies that together constitute a general media system, sources belong to political organisations and other institutions that form the overall political system. The interaction between these two systems gives the content of news media its particular shape. Within these interactions, news media intervene in the political process according to their respective policy positions. Media policy positions in turn derive from the overarching political context of the society. The interrelation between these systems explains the respective newspapers' climate change news policies.

Many scholars agree that news is dominated by only a few types of influential sources. One cannot help but ponder on the implications of such dominance. Perhaps Park's notion of news as knowledge can provide some clue. Park (1940, 2006) outlined a clear distinction between news as "knowledge about" and "acquaintance with." Knowledge deriving from journalism relates to the latter. However, Park also pointed out its inadequacy as, according to him, news "deals ... with isolated events and does not seek to relate them to one another either in the form of causal or in the form of teleological sequences" (1940, p. 675). Because of the nature of knowledge deriving from news, it is not conducive to maintain a sustained interest in significant climate issues internationally (e.g., justice or global responsibility). The demand for climate justice was quite explicit leading up to the Copenhagen conference, despite the fact that the conference was underpinned by conflicting positions among world leaders (e.g., the US vs the BASIC, i.e., Brazil, South Africa, India and China) which led to its failure. However, while the Paris summit was successful, the issue of justice was relatively indistinguishable, not only in the Australian articles but, to some extent, in the Bangladeshi content.

The relatively lower prominence of the "global solidarity or responsibility" theme in the news can also be explained by Beck's politics of invisibility, one of the important elements of global risks discussed in Chapter 4. Here, the evidence suggests that solidarity or responsibility was invisible, not absolutely but relatively, in the examined content. The invisibility of global solidarity was more pronounced in the Australian than in the Bangladeshi articles. The Australian articles also mostly presented the issues of global responsibility and solidarity in a negative manner, for example, questioning the efficacy of international aid. The Bangladeshi articles were vocal in their demand for justice in 2009, but this voice was less strong in 2015, when the writers observed more of a transformative adaptation.

General conclusions: The partial newspapers

While the selection of statements from various sources—political, bureaucratic, expert and activist—confirms the significance of these voices as framing devices of news (Pan & Kosicki 1993) or the construction of perspectives, verification of these statements enables an examination of the extent of variability among different sources in Australia and Bangladesh. However, the professional ideology of balance provides less than satisfactory service to journalism, particularly in environmental news, because in maintaining a so-called balance, the journalists enhanced the climate sceptics' position to a disproportionate extent (Boykoff & Boykoff 2004, 2007). The findings of this study also reveal that the professional ideology of "scrutiny" or "critical gaze" is no "bulwark against bias" in journalism (Lichter et al. 1986, in Schudson & Waisbord 2005, p. 360). In fact, the professional practice of source scrutiny has had little impact on the stance news media adopt in relation to environmental issues. It matters little if a news organisation practises higher or lower quality journalism in terms of scrutiny: its editorial stance is determined through the framing of sources and, ultimately, toes the political line to which it subscribes. This was clearly reflected in the framing (i.e., selection and emphasis) of sources and their juxtaposition in favour of particular positions (e.g., pro-environmental or pro-business) in the construction of news content (see Chapters 5 and 6). This contention is crucial to any understanding of the degree to which news media are inclined to intervene in the internal political processes of various social and political institutions, and to exert "independent influence" on said processes (Schudson 2002, p. 251). This became clear in both of the nation states studied.

A similar pattern in the presence of sources was found in the comparison, which demonstrated a broad similarity of journalistic practice. The point here is neither to undermine the importance of journalistic professional ideology, nor to pronounce news media biased when presenting fewer verified assertions in regard to contentious issues. Rather, a concern of this book is to understand how greater scrutiny of certain sources and less of others in the process of verification engenders the possibility of sustaining certain stances adopted by the respective newspapers. However, some cautionary notes are necessary here. First, this study does not suggest that news media are directly influenced by political strategies and policy decisions (Cook 2005). Rather, it considers the significant impact of organisational influence on the production of news (Gans 2003; Weaver & Loffelholz 2008; Hanitzsch & Mellado 2011, p. 406). This is why this study finds traces of "independent influence" in the coverage of political issues whereby politicians were framed as portraying conflicting sides of an issue, and expert and activist sources were used expediently to legitimise the newspapers' respective concerns about the environment and economy. This kind of influence exerted by the media organisation on its news coverage is increasingly independent from political parties (de Vreese 2001; Strömbäck & Dimitrova 2006). Yet, news media frequently endeavour to intervene in the broader frameworks of a society's social, economic or political structures (van Dijk 2002). And, as this study has shown, where climate change is concerned, this intervention can be subtle as well as blatant. The manifestation of this "independent influence" was found to be far less in the Bangladeshi newspapers than in the Australian publications in terms of verification. This is perhaps because of the Bangladeshi newspapers' pro-environmental orientations, which were aligned with that of the national government, particularly in the case of global climate change.

The important point to note here is the newspapers' different positions: "concerned" and "unconvinced" positions in the Australian content, and "concerned" but with distinctive differences in the Bangladeshi articles (as discussed in Chapter 4). While *The Daily Star* articles viewed Bangladesh as a climatically adaptive nation, the *Prothom Alo* content focused on the scrutiny of various powerful institutions in relation to environmental degradation. Verification of sources mattered less because, irrespective of verification, the newspapers tended to select certain aspects of source statements and evaluate them to endorse their own stances regarding environmental issues.

This is not to imply, however, that verification does not matter at all in the journalistic production process. Rather, verification is practised expeditiously to strengthen certain positions held by particular newspapers. By rehearsing news about climate change and other issues, these continually reiterated positions, although not univocal, enact and maintain relationships between the journalists and their various constituents, e.g., owners, sympathetic political supporters, and, in fact, the majority of their readerships. Through selection and verification, the four newspapers bolster their respective organisations' ideological stances by "defining the range of meaningful, consensually unproblematic information that makes sense to readers without violating the journalists' habitual news-values" (Bell 2013, personal communication). To readers exposed to one newspaper only, such partiality is likely to appear natural and balanced. One of the reasons for this appearance is that, unlike researchers, readers do not always have the advantage of comparing and systematically studying media content, either locally or internationally. However, unilateral application of this conclusion (that newspapers establish their ideological positions while appearing to be professionally sound) could prove problematic. Thus, caution should be observed in extrapolating this assertion to other cases or circumstances. What matters here are not specific positions adopted by the newspapers *per se*, but the diverse processes (such as selection and verification of sources) that journalists endure to reach and empower their readership. Overall, the analysis shows that advanced Western democracies' media systems should not be assumed to serve their readers in more sophisticated ways than their counterparts in the so-called development economies. Newspapers do not "reflect their readers' worlds": they help to maintain or change them. Comparative studies of journalism focused on well-defined, global and national crises show that professionalism and social responsibility may be found in even the most difficult economic environments. This is because news media must always serve two masters—the state (however indirectly) and the people (whom they address directly) or they will fail. Drawing on the evidence of the empirical data, it may be suggested that Bangladeshi newspapers are as sophisticated and successful in these respects as their Australian counterparts.

Post-analysis reflections

The comparison of newspaper content from an economically advanced, industrialised, dispersedly populated "honorary Western country like Australia" (Curran & Park 2000, p. 3) with that of an agriculture-based, densely populated country like Bangladesh has proven quite challenging, albeit remarkably interesting. This comparison of these two vastly dissimilar countries not only identified several types of journalistic practices which deserve further detailed exploration, but also created platforms for further "theoretical sophistication" (Esser & Hanitzsch 2012) in this area of investigation. In journalism studies, comparative research projects are relatively new; one common characteristic is the similarity of cases. West-centrism in the litera-ture is also well-recognised (Josephi 2005), to the extent that Curran and Park (2000) have called for a "de-Westernising" of comparative media studies. Recently, Rodney Tiffen (2013), together with Curran and other colleagues, argued that news practices vary across different national contexts. The find-ings of the current comparative study support these studies and contribute to this debate, which challenges the unthinking generalisation of news practice in the context of advanced globalisation as a taken-for-granted assumption.

This generalisation, however, becomes problematic when journalism as a human institution (Hackett 2017) needs to deal with one of the biggest moral challenges of our time, that is, climate change. The findings of this study show how the practice of journalism becomes trapped in a narrow national logic that raises questions about the relative invisibility of global responsibility in addressing climate change. Perhaps, to face the political and ethical challenges emanating from climate change, the practice of journal-ism as a form of knowledge production requires "North-South dialogues" proposed by Clifford Christians (2015). His recommendation was to shift or orient the central principle of the West-centric journalism from "freedom" and "responsibility" to "accountability." For him,

> To say that agents are accountable for their behaviour means that they can be called to account or judged with respect to their obligations. That is, one can legitimately raise questions or even lay charges, if necessary, and expect reasonable answers.
>
> (p. 48)

In tackling the challenge of climate change, the challenges associated with a journalism that is restricted to the national interest could potentially be addressed by reconsidering the values that underpin the climate debate. However, since the study is not a quantitative examination, its findings are only reflective. It is difficult to make any generalisation about particular types of source use beyond the coverage of environmental issues. More articles should be included in a longitudinal study that would provide a wider variety of principal sources across time and space, enabling a generalising conclusion.

This book has significantly widened the scope for future cross-national research into source selection, verification and framing processes within jour-nalistic practice because it has established diverse national contexts as viable, realistic, conceptually and empirically rewarding study prospects. This widening of scope has the potential to develop a solid framework for com-paring sources and the correlations between particular types of sources and their framing. The latter could also be examined in various media systems, a type of exploration that would provide the opportunity to reassess the notion of news media in the so-called "Third World" as effective development agents, particularly in the rapidly changing media environment in the post-digitisa-tion era (Kleinsteuber 2010). Such scrutiny of journalism in cross-national contexts would benefit future understanding of the relationship between news media organisations and other social institutions in a more fulfilling and meaningful manner.

References

Albaek, E., Christiansen, M. & Togeby, L. 2003, 'Experts in the mass media: Researchers as sources in Danish daily newspapers, 1961–2001', *Journalism & Mass Communication Quarterly*, vol. 80, no. 4, pp. 937–949.

Anderson, A. 1997, *Media, culture and the environment*, Routledge, London.

Bacon, W. 2011, *A sceptical climate: Media coverage of climate change in Australia 2011*, Australian Centre for Independent Journalism, Sydney, NSW.

Beck, U. 1992, *Risk Society*, Sage, London.

Beck, U. 2009, *World at risk*, Polity, Cambridge.

Beck, U. 2010, 'Climate for change or how to create a green modernity', *Theory Culture & Society*, vol. 27, no. 2–3, pp. 254–266.

Becker, H. S. 1967, 'Whose side are we on?', *Social Problems*, vol. 14, no. 3, pp. 239–247.

Bhuiyan, S. 2015, 'Adapting to climate change in Bangladesh: good governance barriers', *South Asia Research*, vol. 35, no. 3, pp. 349–367.

Boyce, T. 2006, 'Journalism and expertise', *Journalism Studies*, vol. 7, no. 6, pp. 889–906.

Boykoff, M. T. 2007, 'Flogging a dead norm? Newspaper coverage of anthropogenic climate change in the United States and United Kingdom from 2003–2006', *Area*, vol. 39, no. 4, pp. 470–481.

Boykoff, M. T. 2011, *Who speaks for climate? Making sense of media reporting on climate change*, Cambridge University Press, Cambridge.

Boykoff, M. T. & Boykoff, J. M. 2004, 'Balance and bias: Global warming and the US prestige press', *Global Environmental Change*, vol. 14, no. 2, pp. 125–136.

Boykoff, M. T. & Boykoff, J. M. 2007, 'Climate change and journalistic norms: A case study of US mass media coverage', *Geoforum*, vol. 38, no. 6, pp. 1190–1204.

Brossard, D., Shanahan, J. & McComas, K. 2004, 'Are issue-cycles culturally constructed? A comparison of French and American coverage of global climate change', *Mass Communication and Society*, vol. 7, no. 3, pp. 359–377.

Caney, S. 2014, 'Two kinds of climate justice: Avoiding harm and sharing burdens', *Journal of Political Philosophy*, vol. 22, no. 2, pp. 125–149.

Carey, J. W. 2007, 'A short history of journalism for journalists: A proposal and an essay', *International Journal of Press/Politics*, vol. 12, no. 1, pp. 3–16.

Carlson, M. 2017, 'Journalism unbound: When professional ethics can no longer hold journalism together', *Journalism & Communication Monographs*, vol. 19, no. 4, pp. 302–306.

Christians, C. 2015, 'North-south dialogues in journalism studies', *African Journalism Studies*, vol. 36, no. 1, pp. 44–50.

Chubb, P. & Bacon, W. 2010, 'Australia: Fiery politics and extreme events', in E. Eide, R. Kunelius & K. Kumpu (eds) *Global climate, local journalism: A transnational study of how media makes sense of climate summits*, Projekt Verlag, Freiberg, Germany, pp.51–65.

Climate Council of Australia 2015, 'Half way to Paris: How the world is tracking on climate change', accessedMay 10, 2017 available: https://www.climatecouncil.org.au/uploads/7ec258783dd2367efa806f6dc6c9a54a.pdf.

Cook, T. E. 2005, *Governing with the news: The news media as a political institution*, 2nd edn, University of Chicago Press, Chicago, IL.

Cottle, S. 2009, *Global crisis reporting: Journalism in the global age*, Open University Press, New York.

Curran, J. 2002, *Power without responsibility: The press and broadcasting in Britain*, Routledge, London.

Curran, J. & Park, M. J. 2000, *De-Westernizing media studies*, Routledge, London.

de Vreese, C. H. 2001, 'Europe in the news: A cross-national comparative study of the news coverage of key EU events', *European Union Politics*, vol. 2, no. 3, pp. 283–307.

Downs, A. 1972, 'Up and down: The issue-attention cycle', *The Public Interest*, vol. 28, Summer, pp. 38–51.

Driessen, P. 2002, *Eco-imperialism: Green power, black death*, Merrill Press, Bellevue, WA.

Eide, E., Kunelius, R. & Kumpu, K. 2009, 'Blame, domestication and elite perspectives in global media climate: Limits of transnational professionalism in journalism', paper presented at the Global Dialogue Conference, Arhus, Denmark, November 4–6.

Eide, E. & Ytterstad, A. 2011, 'The tainted hero: Frames of domestication in Norwegian press representation of the Bali climate summit', *International Journal of Press/Politics*, vol. 16, no. 1, pp. 50–74.

Entman, R. M., Matthews, J. & Pellicano, L. 2009, 'Framing politics in the news: Nature, sources and effects', in K. Wahl-Jorgensen & T. Hanitzsch (eds) *The handbook of journalism studies*, Routledge, London, pp. 175–190.

Ericson, R. V., Baranek, P. M. & Chan, J. B. L. 1989, *Negotiating control: A study of news sources*, Open University Press, Milton Keynes.

Esser, F. & Hanitzsch, T. 2012, 'On the why and how of comparative inquiry in communication studies', in F. Esser & T. Hanitzsch (eds) *The handbook of comparative communication research*, Routledge, New York, pp. 3–24.

Gans, H. J. 2003, *Democracy and the news*, Oxford University Press, New York.

Giddens, A. 2011, *The politics of climate change*, Polity, Cambridge.

Golding, P. 1977, 'Media professionalism in the third world: The transfer of an ideology', in J. Curran, M. Gurevitic & J. Woollacott (eds) *Mass communication and society*, Arnold, London, pp. 291–308.

Hackett, A. R. 2017, 'Democracy, climate crisis and journalism: Normative touchstones', in A. R. Hackett, S. Forde & K. Foxwell-Norton (eds) *Journalism and climate crisis: Public engagement, media alternatives*, Routledge, London, pp. 20–48.

Hall, S., Critcher, C., Jefferson, T., Clarke, J. & Roberts, B. 1978, *Policing the crisis: Mugging, the state, and law and order*, Macmillan, London.

Hanitzsch, T. & Mellado, C. 2011, 'What shapes the news around the world? How journalists in eighteen countries perceive influences on their work', *International Journal of Press/Politics*, vol. 16, no. 3, pp. 404–426.

Hannigan, J. 2006, *Environmental sociology*, Routledge, London.

Hansen, A. 1991, 'The media and the social construction of environment', *Media, Culture & Society*, vol. 13, no. 4, pp. 443–458.

Harvey, D. 1996, *Justice, nature and the geography of difference*, Blackwell, Oxford.

Hulme, M. 2009, *Why we disagree about climate change: Understanding controversy, inaction and opportunity*, Cambridge University Press, Cambridge.

Huq, S. 2019, 'Changing the climate narrative: time for transformational adaptation for climate resilient Bangladesh', *The Daily Star*, January 2, accessed February 1,

2019, available: https://www.thedailystar.net/opinion/politics-climate-change/news/changing-the-climate-narrative-1681573.

Josephi, B. 2005, 'Journalism in the global age: Between normative and empirical', *Gazette: The International Journal for Communication Studies*, vol. 67, no. 6, pp. 575–590.

Kleinsteuber, J. H. 2010, 'Comparing west and east: A comparative approach of transformation', in B. Dobek-Ostrowska, M. Glowacki, K. Jakubowicz, & M. Sukosd (eds) *Comparative media systems: European and global perspectives*, Central European University Press, Budapest, New York, pp. 23–40.

Kovach, B. & Rosenstiel, T. 2007, *The elements of journalism: What news people should know and public should expect*, Three Rivers Press, New York.

Laksa, U. 2014, 'National discussions, global repercussions: ethics in British newspaper coverage of global climate negotiations', Environmental Communication, vol. 8, no. 3, pp. 368–387.

Lichter, S. R., Rothman, S. & Lichter, L. S. 1986, *The media elite: America's new powerbroker*, Adler & Adler, Bethesda, MD.

Manne, R. 2011, 'Bad news: Murdoch's Australian and the shaping of the nation', *Quarterly Essay*, vol. 43, September, pp. 1–119.

McGaurr, L. & Lester, L. 2013, 'Country studies: Australia', in J. Painter (ed.) *Climate change in the media: Reporting risk and uncertainty*, University of Oxford, London, pp. 79–88.

McKnight, D. 2010, 'A change in the climate? The journalism of opinion at news corporation', *Journalism*, vol. 11, no. 6, pp. 693–706.

Nash, C., Chubb, P. & Birnbauer, B. 2009, 'Fighting over fires: Climate change and the Victorian bushfires of 2009', paper presented at the Global Dialogue Conference, Aarhus, Denmark, November 3–6.

Painter, J. 2013, *Climate change in the media*, I. B. Taurus, London.

Pan, Z. & Kosicki, G. 1993, 'Framing analysis: An approach to news discourse', *Political Communication*, vol. 10, no. 1, pp. 55–75.

Park, E. R. 1940, 'News as a form of knowledge: A chapter in the sociology of knowledge', *American Journal of Sociology*, vol. 45, no. 5, pp. 669–686.

Park, E. R. 2006 (1940), 'News as a form of knowledge', in S. G. Adam & R. P. Clarke (eds) *Journalism: The democratic craft*, Oxford University Press, Oxford.

Raupach, M. R., Canadell, J. G. & Le Quéré, C. 2008, 'Anthropogenic and biophysical contributions to increasing atmospheric CO_2 growth rate and airborne fraction', *Biogeosciences*, vol. 5, pp. 1601–1613.

Roberts, J. T. & Parks, B. 2007, *A climate of injustice: Global inequality, north-south politics, and climate policy*, MIT Press, Boston, MA.

Schudson, M. 2002, 'The news media as political institutions', *Annual Review of Political Science*, vol. 5, pp. 249–269.

Schudson, M. & Waisbord, S. 2005, 'Toward a political sociology of the news media', in T. Janoski, R. R.Alford, A. M.Hicks & M. A. Schwartz (eds), *The handbook of political sociology: States, civil societies and globalization*, Cambridge University Press, Cambridge, pp. 350–366.

Shanahan, M. 2006, 'Science journalism: Fighting a reporting battle', *Nature*, no. 443, pp. 392–393.

Shanahan, M. 2011, 'Time to adapt? Media coverage of climate change in non-industrialised countries', in T. Boyce & J. Lewis (eds) *Climate change and the media*, Peter Lang, Frankfurt, New York, pp. 145–157.

Snow, D. A. & Benford, R. D. 2000, 'Clarifying the relationship between framing and ideology in the study of social movements: A comment on Oliver and Johnston', *Mobilization*, vol. 5 no. 2, pp. 55–60.

Sovacool, B. K. 2017, 'Bamboo eating bandits: Conflict, inequality and vulnerability in the political ecology of climate change adaptation in Bangladesh', *World Development*, vol. 102, pp. 183–194.

Strömbäck, J., & Dimitrova, D. V. 2006, 'Political and media systems matter: A comparison of election news coverage in Sweden and the United States', *The International Journal of Press/Politics*, vol. 11, no. 4, pp. 131–147.

Tiffen, R., Jones, K. P., Rowe, D., Aalberg, T., Coen, S., Curran, J., Hayashi, K., Iyengar, S., Mazzoleni, G., Papathanassopoulos, S., Rojas, H. & Soroka, S. 2013, 'Sources in the news', *Journalism Studies*, vol. 15, no. 4, pp. 1–19.

Tuchman, G. 1978, *Making news: A study in the construction of reality*, Free Press, New York.

van Dijk, T. 2002, 'Political discourse and political cognition', in P. Chilton & C. Schaffner (eds) *Politics as text and talk: Analytic approaches to political discourse*, John Benjamins, Amsterdam, pp. 203–223.

Weaver, D. & Loffelholz, M. 2008, 'Questioning national, cultural and disciplinary boundaries: A call for global journalism research', in M. Löffelholz & D. Weaver (eds) *Global journalism research: Theories, methods, findings, future*, Blackwell, Oxford, pp. 3–12.

Wimmer, J. & Quandt, T. 2006, 'Living in the risk society', *Journalism Studies*, vol. 7, no. 2, pp. 336–347.

Appendix 1: Analysis of editorial commentaries in Chapter 4

For the purpose of discourse analysis described in Chapter 4, I adopted van Dijk's (1993, p. 264) persuasive strategies and developed the coding categories to analyse the selected editorial commentaries (total 80: 20 each for 2009 and 2015 from each of the four newspapers). The selection of editorial commentaries was based on Seale's (2012) purposive sampling, and followed by Lindlof and Taylor's (2011) coding process. In this sampling, an article was chosen on the basis of its relevance to the climate change topics. Below are a few examples from newspaper articles to illustrate the coding.

Table A1.1 Categories and examples: Australian newspapers

Categories	Elements	Examples[a]
Argumentation	The negative evaluation of an events or topic follows from facts	Is Copenhagen just an exercise in wealth redistribution?
Rhetorical figure	Hyperbolic enhancement of "their" negative actions and "our" positive action	Derailed by efforts to end world poverty and redistribute wealth under the guise of helping the developing world adapt to climate change. The negotiating process has been distorted by NGOs and developing nations with a cargo cult mentality who see climate change as a way to attract aid dollars from the West
Lexical style	Choice of words that imply positive or negative evaluation	Symbolism triumphed over substance yesterday with the decision to exclude clean coal from funding projects designed to develop this technology
Storytelling	Giving plausible details above negative features of the events	The shemozzle in the Danish capital has done little to encourage voters' trust in the global efforts to get to grips with climate change

Categories	Elements	Examples[a]
Structural emphasis	Emphasis of the negative actions in headline and lead summaries	Headline: The parallel universe with a life of its own
Quotation	Quoting credible witness, sources or experts	Dr Plimer is an Adelaide geologist who argues that rather than taking a 150-year time span to assess the problems of global warming, we need to look back several million years

a Australian (2009).

Table A1.2 Categories and examples: Bangladeshi newspapers

Categories	Elements	Examples[a]
Argumentation	The negative evaluation of an events or topic follows from facts	The fact of the matter is that greenhouse gas emissions have kept rising. In similar manner, temperatures too have been going up and it is feared will rise to above 3.2 degrees Celsius. The ball is now clearly in the court of the developed nations to live up to their historical responsibility to the disadvantaged nations to cut back on greenhouse emissions and help them adapt to climate change
Rhetorical figure	Hyperbolic enhancement of their negative actions and our positive action	A draft "climate change agreement" leaked to the media in Copenhagen gives "more power to rich countries and sidelines the United Nations" role in all future climate change agreements," The Guardian reported. This must not be the case. The conference is already fraught with many misgivings and differences … Yet the participating parties must be hoping against hope for a conclusion that matters most to the vulnerable
Lexical style	Choice of words that imply positive or negative evaluation	Tuvalu is threatened with extinction and so is the Maldives. One-fifth of Bangladesh will be gone, thus increasing the pressure on a country already engaged in several battles at the socio-economic levels
Storytelling	Giving plausible details, negative features of the events	Just last May, millions were displaced when Cyclone Aila hit the low-lying country
Structural emphasis	Emphasis of the negative actions in headline and lead summaries	Will the world listen to this boy of Bangladesh to cut CO_2 emission?

Categories	Elements	Examples[a]
Quotation	Quoting credible witness sources or experts	Germanwatch concludes its Global Climate Risk Index 2010 ahead of the Copenhagen climate change conference. highlighting, "In countries like Bangladesh, extreme events have become a constant danger"

a Daily Star (2009).

References

Australian, The 2009, 'The parallel universe with a life of its own', December 17.
Daily Star, The 2009, 'Can Copenhagen deliver "hope" for Bangladesh?', December 18.
Lindlof, R. T. & Taylor, C. B. 2011, *Qualitative communication research methods*, 3rd ed., Sage, London.
Seale, C. 2012, *Researching society and culture*, Sage, London.
van Dijk, T. 1993, 'Principles of critical discourse analysis', *Discourse & Society*, vol. 4, no. 2, pp. 249–283.

Appendix 2: Description: Data collection

Altogether, the four newspapers published 3,998 articles containing one or more search terms pertinent to climate change during the study periods in 2009 and 2015. Of them, the two Australian newspapers published 3,180 articles, while the Bangladeshi newspapers published the other 818 articles. However, not all Australian articles were highly relevant to the study because the Factiva database returned every article with just a mere mention of any of the search terms. To obtain a precise picture of the coverage, less relevant articles were excluded from analysis. Following this exclusion, a total of 3,062 articles were found suitable for analysis. Of them, 1,683 articles were from 2009 and 561 from 2015.

However, the same did not occur in the case of the Bangladeshi content because these articles were collected manually from the archives of the respective newspaper websites as they were not available in any database. The following is a year-wise breakdown of the selected articles for analysis.

Table A2.1 Number of selected articles from Australia and Bangladesh, 2009 and 2015

Country	Articles (2009)	Articles (2015)	Total
Australia	1,683	561	2,244
Bangladesh	543	275	818
Total	2,226	836	3,062

Table A2.2 Climate change article type, Australia

Article type	2009		2015	
	Australian	*SMH*	*Australian*	*SMH*
News	430 (41.43%)	300 (46.51%)	151 (49.35%)	149 (58.43%)
Feature	105 (10.11%)	83 (12.87%)	2 (1.32%)	8 (3.14%)
Editorial	61 (5.88%)	31 (4.81%)	11 (3.59%)	12 (4.70%)
Commentary	356 (34.30%)	180 (27.90%)	106 (34.64%)	68 (26.67%)
Other	86 (8.29%)	51 (7.90%)	36 (11.76%)	18 (7.05%)
Total	1,038	645	306	255

SMH = Sydney Morning Herald.

Table A2.3 Climate change article type, Bangladesh

Article type	2009		2015	
	Daily Star	*Prothom Alo*	*Daily Star*	*Prothom Alo*
News	180 (58.44%)	133 (56.60%)	86 (57.71%)	71 (56.35%)
Feature	7 (2.27%)	30 (12.76%)	6 (4.03%)	16 (12.70%)
Editorial	11 (3.57%)	6 (2.55%)	8 (5.37%)	4 (3.17%)
Commentary	64 (20.78%)	44 (18.72%)	33 (22.15%)	23 (18.25%)
Other	46 (14.93%)	22 (9.36%)	16 (10.74%)	12 (9.52%)
Total	308	235	149	126

Description of different sources

Principal source definition: Principal sources are individual news sources (e.g. politicians, bureaucrats, experts or citizens) who provide important information/statements in support of or against the main theme of a published article.

Values	Source type	Source description
1	Political	Political sources include individual politicians who make assertion/s in support of or against the main theme of the article
2	Business	The term "business sources" allies to individuals who have experienced about some aspects of the economic issues, and made statement/s supporting or opposing the main theme of the article
3	Bureaucratic	Bureaucratic sources include official/s (e.g., bureaucrats) from different government departments or other organisations who make statement/s in support of or against the main theme of the article
4	Expert/ scientific	Expert sources include individual experts (academics, scientists) on river, water or climate issues, who make statement/s supporting or opposing the main theme of the article
5	Activist	Activist source are individual activists (e.g., non-government organisations working for climate change or aboriginal activists attempting to protect their land rights) who make statement/s supporting or opposing the main theme of the article
6	Citizen	Citizen or lay sources include individuals who have experienced or witnessed some aspects of the issues pertaining to the item and make statement/s supporting or opposing the main theme of the article
7	Other	Other sources include individuals who make statements supporting or opposing the main theme of the article, but do not fall into any of the above categories (e.g., anonymous sources)

Verification of statements

Verification results when the journalist/author of an article has checked the veracity of different source statement/s with other sources or logical argumentations. Whether an entire article is adequately verified or not is determined by the number of statement/s verified in this manner.

Values:

1   1 = few statements about climate change issues are checked
2   2 = no statements about climate change issues are checked

Table A2.4 Sources in Australian climate change news

Source type	2009		2015		Total
	Australian	SMH	Australian	SMH	
Political	367 (48.74%)	239 (51.50%)	167 (41.43%)	115 (34.84%)	888
Bureaucratic	98 (13.01%)	62 (13.36%)	26 (6.45%)	29 (8.78%)	215
Expert	126 (16.73%)	77 (16.59%)	62 (15.38%)	66 (20%)	331
Activist	33 (4.38%)	28 (6.03%)	35 (8.68%)	32 (9.69%)	128
Business	76 (10.09%)	29 (6.25%)	72 (17.86%)	55 (16.66%)	232
Citizen	12 (1.59%)	0 (0%)	3 (0.75%)	12 (3.63%)	27
Other	41 (5.44%)	29 (6.25%)	38 (9.42%)	21 (6.36%)	129
Total	753	464	403	330	1,950

Table A2.5 Principal sources in Australian climate change news

Source type	2009		2015	
	Australian	SMH	Australian	SMH
Political	116 (44.78%)	84 (54.54%)	71 (50.35%)	45 (36.58%)
Bureaucratic	37 (14.28%)	15 (9.74%)	10 (7.09%)	12 (9.75%)
Expert	37 (14.28%)	30 (19.48%)	23 (16.31%)	26 (21.13%)
Activist	15 (5.79%)	7 (4.54%)	9 (6.38%)	12 (9.75%)
Business	40 (15.44%)	12 (7.79%)	21 (14.89%)	24 (19.51%)
Citizen	0	0	0	0
Other	14 (5.40%)	6 (3.89%)	7 (4.96%)	4 (3.25%)
Total	259	154	141	123

Table A2.6 Verified sources in Australian articles

Source type	2009				2015			
	Australian		SMH		Australian		SMH	
	Total (T)	Verified (V)	(T)	(V)	(T)	(V)	(T)	(V)
Political	116	40 (34.5%)	84	30 (35.7%)	71	33 (46.5%)	46	23 (50%)
Bureaucrats	31	10 (34.4%)	13	5 (38.4%)	10	5 (50%)	12	8 (66.7%)
Experts	42 (65.38%)	23 (54.7%)	32	11 (34.4%)	23	14 (63.6%)	26	17
Activists	15	7 (46.6%)	7	3 (42.8%)	9	6 (66.6%)	12	7 (58.3%)
Business	40	16 (40%)	12	6 (50%)	21	8 (38.1%)	24	11 (45.8%)
Citizen	0	0	0	0	0	0	0	0
Other	0	0	0	0	0	0	0	0
Total	244	96	148	55	134	66	120	66

Table A2.7 Sources in Bangladeshi climate change news

Source type	2009		2015		Total
	Daily Star	Prothom Alo	Daily Star	Prothom Alo	
Political	89 (42.99%)	102 (40.16%)	62 (30.09%)	49 (24.62%)	302
Bureaucratic	46 (22.22%)	52 (20.47%)	49 (23.78%)	55 (27.63%)	202
Expert	28 (13.52%)	45 (17.71%)	21 (10.19%)	28 (14.07%)	122
Activist	22 (10.63%)	19 (7.48%)	30 (14.56%)	26 (13.07%)	97
Business	4 (1.93%)	8 (3.15%)	18 (8.73%)	3 (1.50%)	33
Citizen	5 (2.41%)	11 (4.33%)	7 (3.39%)	14 (7%)	37
Other	13 (6.28%)	17 (6.69%)	19 (9.22%)	29 (14.57%)	78
Total	207	254	206	204	871

Table A2.8 Principal sources in Bangladeshi climate change news

Principal sources	2009		2015	
	Daily Star	Prothom Alo	Daily Star	Prothom Alo
Political	41 (49.39%)	60 (43.16%)	31 (36.04%)	25 (30.86%)
Bureaucratic	16 (19.27%)	13 (9.35%)	16 (18.60%)	18 (22.22%)
Expert	13 (15.66%)	15 (10.79%)	12 (13.95%)	11 (13.58%)
Business	2 (2.40%)	3 (2.15%)	9 (10.46%)	1 (1.24%)
Activist/NGO	8 (9.63%)	16 (11.51%)	7 (8.13%)	19 (13.58%)
Citizen	0 (0.0%)	0 (0.0%)	2 (2.32%)	3 (3.70%)
Other	3 (3.61%)	32 (23.02%)	9 (10.46%)	12 (14.81%)
Total	83	139	86	81

Table A2.9 Verified sources in Bangladeshi climate change news

Source type	2009				2015			
	Daily Star		Prothom Alo		Daily Star		Prothom Alo	
	Total (T)	Verified (V)	(T)	(V)	(T)	(V)	(T)	(V)
Political	41	5 (12.20%)	60	7 (11.66%)	31	2 (6.45%)	16	9 (56.25%)
Bureaucratic	16	3 (18.75%)	13	5 (38.46%)	16	5 (31.25%)	18	13 (72.22%)
Expert	13	4 (30.76%)	15	6 (40%)	16	3 (18.75%)	19	11 (57.89%)
Activist	8	0 (0.0%)	16	0 (0.0%)	6	3 (50%)	19	11 (57.89%)
Business	2	0 (0.0%)	3	0 (0.0%)	9	1 (11.1%)	1	0 (0.0%)
Citizen	8	3 (37.5%)	16	9 (56.25%)	5	2 (40%)	4	3 (75%)
Other	0	0	0	0	0	0	0	0
Total	88	15	123	27	83	16	77	47

Index

Taylor & Francis eBooks

www.taylorfrancis.com

A single destination for eBooks from Taylor & Francis
with increased functionality and an improved user
experience to meet the needs of our customers.

90,000+ eBooks of award-winning academic content in
Humanities, Social Science, Science, Technology, Engineering,
and Medical written by a global network of editors and authors.

TAYLOR & FRANCIS EBOOKS OFFERS:

A streamlined
experience for
our library
customers

A single point
of discovery
for all of our
eBook content

Improved
search and
discovery of
content at both
book and
chapter level

REQUEST A FREE TRIAL
support@taylorfrancis.com

 Routledge
Taylor & Francis Group

 CRC Press
Taylor & Francis Group

Milton Keynes UK
Ingram Content Group UK Ltd.
UKHW040053071024
449327UK00019B/526